ORGANIZATION
AFTER
SOCIAL MEDIA

Geert Lovink and Ned Rossiter

Organization after Social Media
Geert Lovink and Ned Rossiter

ISBN 978-1-57027-338-4

Cover concept and design by Amir Husak
Interior layout by Margaret Killjoy

Released by Minor Compositions 2018
Colchester / New York / Port Watson

Minor Compositions is a series of interventions & provocations drawing from
autonomous politics, avant-garde aesthetics, and the revolutions of everyday life.

Minor Compositions is an imprint of Autonomedia
www.minorcompositions.info | minorcompositions@gmail.com

Distributed by Autonomedia
PO Box 568 Williamsburgh Station
Brooklyn, NY 11211

www.autonomedia.org
info@autonomedia.org

CONTENTS

ACKNOWLEDGEMENTS

OUR COLLABORATIVE WORK ON THE CONCEPT OF ORGANIZED NET-works has grown organically over a period of more than a decade. Time and again, we published essays on the topic, amazed at how smoothly the texts came into being; and how these two magic words linked together, attracted materials, experiences, and ideas. From the very beginning of our friendship, back in Melbourne in 2001, when we contributed to the collective formation of the Australian Fibreculture network, our writing together proved to be an effortless adventure. It's hard to say if the time is right to bring all the material together in a book. We felt the urgency from early on but did not feel we were in a rush.

The references in the back document the date of first versions and publication sites of the texts collected in this book. A couple of chapters aside, we have chosen not to radically regroup the material. Instead, we shortened the texts, taken out the obligatory explanations of what organized network are and can be, added some new writings, and updated the insights here and there.

Texts were written in Melbourne, Amsterdam, Berlin, Shanghai, and Sydney.

Along the way, we have benefitted from the welcome invitations and generous contributions by editors and copy-editors, all of whom have improved the clarity of ideas and arguments. In no particular order, we thank Morgan Currie, Anna Watkins Fisher, Eric Kluitenberg and David Garcia, Graham Miekle, Rosi Braidotti and Maria Hlavajova, Lina Dencik and Oliver Leistert, Steirischer Herbst and Florian Malzacher, Kelly Gates, Geraldine Barlow, Rasa Smite, Wendy Hui Kyong Chun and Thomas Keenan, Mark Deuze, Marco Berlinguer, and Hilary Wainwright.

The concept of organized networks has been developed in dialogue and texts written together with Soenke Zehle, Brett Neilson, Ippolita, and Gabriella Coleman and discussions with Andrew Murphie, Anna Munster, Akseli Virtanen, Jon Solomon, Sandro Mezzadra, Paolo Do, and Gigi Roggero. Like so many, we have been organizing our own networks for years. In the Beijing summer of 2007, 'orgnets' came together with Mónica Carriço, Bert de Muynck, and those who participated in the counter-mapping project there. In 2009, we organized Winter Camp at the Institute of Network Cultures in Amsterdam along with twelve participating networks. Both of these projects have informed our thinking in this book, and we thank all those involved. There have been numerous invitations to talk about the concept and project of organized networks at workshops, seminars, conferences, activist meetings, and lectures. We thank all the organizers of those events.

Thanks also to Stevphen Shukaitis and Autonomedia for coming on board with this project.

Thanks to the support of Linda and Kazimir, and Maren and Willem.

Amsterdam/Sydney, November 2017

1.

INTRODUCTION

One day, in retrospect, the years of struggle will strike
you as the most beautiful.
— Sigmund Freud

THESE DAYS, STRATEGIC CONSIDERATIONS FOR POLITICAL ORGANIZA-
tion no longer bother with mediation, representation, and identity
politics. Instead, the key question revolves around the design of new
(sustainable) organizational forms. What is the social today, if not
social media? It is not enough to indulge in the aesthetics of revolt.
Flaws in the 19th and 20th century models of the party, the union,
and the movement are easy to detect, but what's replacing them? It is
tempting to say that the network is the dominant form of the social: a
programmed life under permanent surveillance. What can replace the
corporate walled gardens such as Facebook and Twitter? Our answer
to this question is a firm and open one: a federation of organized networks, sustainable cells that operate as secret societies.

Many have already identified social networks as a conspiratorial
neoliberal invention that, in the end, only benefits the global elite.
Think of the vampire data mining economies made possible with all

your searches, status updates, likes, etc. The algorithmic modulation of networks generates patterns of data that hold economic value for social media corporations and finance capital. These extraction machines produce a subject Maurizio Lazzarato calls "indebted man." Exodus for the multitudes, it would seem, is a futile proposition.

Nearly twenty years into the 21st century we can conclude that global elites are not threatened by temporary uprisings and will only be questioned by an offensive counter-power that is capable of learning and incorporating its own trial-and-error experiments of daily struggles into the social body. But wait a minute, how does this intersect with the technological condition? Digital networks have been discredited for their short-lived character that merely reproduce the hegemonic fragmentation of desperate subjects. No matter how legitimate such structural proposals are, they often end up in a retromania of social imagination.

In defense of the network. Fatigue has well truly and set in. Time has been stolen. Sleep has been injured (Jonathan Crary). Online efforts have been exploited to the max by the cynical social media and their economies of data mining. The network form has either eroded or been totally expropriated and relocated to the cloud. The shift from networks to cloud-based media has been a setback, a regressive move. People are tired of updating and maintaining the labor of online administration. The work of securing social capital is now a chore preferably outsourced to PAs on the global peripheries. If, as an influencer, you don't have the resources to hire your personal Tweeter, then you have to carve out the time in the day to shoot your own selfies. Migrating across platforms has now become part of many people's digital biographies. The tedium of doing this repeatedly has well and truly set in. Will young people be the first among those to terminate the contract with social media?

So what to do, and where to go in order to live and work in ways autonomous from these technologies of capture? One place to start is at the level of organization, which requires addressing the problematic of infrastructure. Our proposition is that the (legitimized) desire to build lasting collective forms should grow out of 21st century materialities and not be based on nostalgic notions of mass organization. Instead of dismissing the network as such, we propose to rewire, recode, and redefine its core values and develop new protocols for the social, which, in today's society, is technical in nature.

Today's problem is no longer the Art of Mobilization. Organized networks have access to an array of tools, though a relatively limited range of social media platforms are more often the preferred choice for mass mobilization. Memes spread like wildfire in real-time. We know how to put together campaigns, create shit storms, and go viral: read the fucking manual, as hackers in the past used to say. Majorities are enraged and rally against climate change, repression, violence, rape, authoritarian rule, education cuts, poverty, and job losses. We sign petitions and maybe even shut down a website. But we need to shift these technical practices to another level.

Designing encryption as a standard is one core technical practice relevant to organized networks that we see developing post-Snowden and the National Security Agency (NSA) revelations. Encryption accessible on a mass scale is an example of an alternative at work, of the time-old paradox of constraints creating possibility. Pre-Snowden, encryption was for a handful hackers, high government communications, and corporate transactions with something at stake. But we are now are in the midst of a tipping point where individual users – and less so organizations – are deciding to encrypt communications. So the next level would be to see more coordinated efforts at encrypting collective communication.

Is encryption an example of standards scaling up? A form of civil defense in a time of serious technological onslaught? What can people do to protect the privacy of communication and the dignity of their online life? Of course forms of secure communication goes on within social and political movements among the chief organizers or facilitators. But less so across the social base of the movements who are not so much involved in decision making. This leads to potential dead-end streets in the forms of content and organization. What is the broader potential of crypto?

The mass introduction of cryptography is a reassessment of the secret society as a cultural technique. Invisible and secret organizations have been accused of the "terror of the informal," which is reprimanded for not being accountable. This politically correct rhetoric needs to be countered with the argument that organized networks are not public organizations or state bodies. The trick is to achieve a form of collective invisibility without having to reconstitute authority. Organized networks are not vanguard parties. The party in its original sense claims to articulate the general interest and will of the people. As an organizational form, the party is a sustainable structure that is here to stay regardless of its own fluctuations in the polls. But the party today is without passion and holds little relevance to people's daily social lives and communication practices.

The secret society has always been connected to conspiracy, but what if it becomes not only a necessity but a civil duty? Many of the other possible alternatives lead to the romantic world of offline. Think "maker cultures" – which can't function anyway without the marketing power of social media and the distribution and production systems of global supply chains. The slow food movement is another example, which is now thoroughly commercialized as well. Forget the nostalgia option. Offline romanticism is also part of the NSA repertoire when they break into your house: this is the exception in their weapons armory, and why they invest so much in online surveillance and hardware manipulation.

The social-technological default of encryption makes secret societies mainstream. The question of what issues or agendas to pursue remains open and undecided. Encrypted communication requires a motivating cause. Once this is identified, networks could begin to organize in more secure and sustainable ways.

Organization under Platform Capitalism

In an age of algorithmic governance and preemptive action, the prevailing schema of politics is orchestrated around data analytics of social media. Politicians gravitate toward Facebook and Twitter on the advice of their minders, assuming the pulse of the masses can be aggregated and calibrated back into policy settings. Oversight of this cybernetic machine is also pursued by humanities and social science researchers invested in digital methods that index the inputs of civil society in participatory mode. Against this managerial model of governance and knowledge production, the question of correspondence between data and the world of objects and things remains elusive as long as schemas of intelligibility command institutional, epistemological, and political hegemony.

The fantasy of government through cybernetics was trialed at the prototype level in Stafford Beers' experiments in data-driven socialism in Allende's Chile in the 1970s.[1] Such a model was revived in recent

---

1   See Eden Medina, *Cybernetic Revolutionaries: Technology and Politics in Allende's Chile* (Cambridge, Mass.: MIT Press, 2011).

years with the attempt by the P2P Foundation, along with initiatives such as Bernard Stiegler's L'Institut de recherche et d'innovation (IRI), to install peer-to-peer models of socio-economic production and education in Ecuador. The attempt to implement a counter-hegemonic system in this instance failed primarily because of a struggle to find a common language. This is not a problem of what Naoki Sakai terms "homolingual translation" so much as a problem of making a concept quantitatively jump into the form of a meme that penetrates and infects institutional mentalities.[2]

As much as the free software and creative commons movements have hit the mainstream they have paradoxically remained in the margins of the power of the stacks, otherwise known as platform capitalism. In earlier times there was either the mainstream or the margin. You could exist in one but not both. Within a near universal condition of a mainstream without margins, the capacity to devise and unleash the power of critique is consigned to the *Trauerspiel* of modernity. Immanence without an outside is submission with occasional resistance whose only effect is to supply data-driven capitalism with a surplus of records and related metatags.

For all the attempts to establish a critical mass for alternative practices in the age of the Anthropocene, which manifest as networks of organic food suppliers, hipster maker economies, co-working spaces, urban gardening, and renewable energies, there remains a dependency on mainstream architectures from global logistics to data centers and the perpetuation of an international division of labor. There is no visible prospect of these core planetary systems being overhauled or replaced. Despite the proliferation of these sort of alternative practices, the decline in global working standards and employment opportunities is inseparable from the penetrative force of finance capitalism.

However much the possibility of thinking the Hegelian totality remains as a utopian position from which to overcome the fragmentation and dissipation of material and social life, the digital architectures that operationalize the world increasingly withdraw from the grasp of the human. Even those such as Yanis Varoufakis, who have glimpsed

---

2   On the distinction between "homolingual" and "heterolingual" translation, see Naoki Sakai, *Translation and Subjectivity: On "Japan" and Cultural Nationalism* (Minneapolis: University of Minnesota Press, 1997). See also Naoki Sakai, "Translation," *Theory, Culture & Society* 23.2-3 (2006): 71–86.

the inner-workings of the Euro-technocratic elite, are unable to manifest proposals for a movement of the disaffected. The network imaginary cannot on its own perform the work of implementation. Why? Because the stacks reign supreme.

The consolidation of resignation is one option. The now struggling agenda of the Mont Pèlerin Society is another. Regional geopolitical giants of Putin's Russia or the Beijing Consensus may, for all we know, deliver the path to restoration for a "multi-polar" future able to withstand the ravages of capitalism in ways not reliant on Silicon Valley's engineering logic of techno-solutionism. But unless we wish to commit to a paternalistic vision to be realized by whatever geopolitical elite invested in the global redistribution of wealth and resources, the question of organization without state-enmeshed sovereignty remains to be addressed.

Organization aimed at clutching power from above will do nothing in terms of forging a global grammar able to design concepts that critique and direct debates on issues and conditions in order to regain the initiative. Cognitive capitalism obtains power, in part, because of its binding capacity.[3] It is able to distribute and implement a coherent message across a vast range of institutional and organizational settings. In other words, cognitive capitalism holds an elective affinity with technologies of mediation. Without continuous network maintenance, it falls apart. Rituals of organization are required to galvanize sociality in coherent rather than perpetually dispersed forms and practice.[4]

Where are the forms of organization that regenerate the collective confidence that typified the historical avant-garde? Can new modes of organization function in a centrifugal manner to escape the sectarianism of the group dynamic? A decade ago we proposed the concept of organized networks as a new institutional form in response to the "walled gardens" of social media. We foregrounded the need for a strategic turn that could address the problem of sustainability of social organization. Neighboring concepts such as "platform cooperativism"

---

3   See Yann Moulier Boutang, *Cognitive Capitalism*, trans. Ed Emery
     (Cambridge: Polity Press, 2011).

4   See James Carey, "A Cultural Approach to Communication," in
     Communication as Culture: Essays on Media and Society (New York and
     London: Routledge, 1992), 13–36.

and the many experiments in social centers and educational infrastructures such as "freethought" are strong examples of how the work of invention is manifesting as new organizational forms.[5]

A distributed laboratory of thought is needed that fuses intellectual and political invention without the clientelism of the think tank model. A praxis that dispenses with the misguided sentiment of post-capitalist economies and all the privilege that entails. The inquiry of this book contributes to a wider intellectual, political, and artistic cataloguing of concepts, problems, and conditions that experiment with the organization of thought not consigned to the affirmation of the transcendent. How to unleash concepts that organize totality as a distributed and differentiated architecture is key to the formation of autonomous infrastructures able to withstand the monopoly on decision gifted to algorithmic capitalism.

From Weak Ties to Strong Links

Sloganism: "I feel protected by unpublished Suite A algorithms." (J. Sjerpstra) – "I am on an angry squirrel's shitlist." – Join the Object Oriented People – "When philosophy sucks—but you don't." – "See you in the Sinkhole of Stupid, at 5 pm." – "I got my dating site profile rewritten by a ghost writer." – "Meet the co-editor of the Idiocracy Constitution" – The Military-Entrepreneurial Complex: "They are bad enough to do it, but are they mad enough?" – "There really should be something like Anti-Kickstarter for the things you'd be willing to pay to have not happen." (Gerry Canavan) – Waning of the Social Media: Ruin Aesthetics in Peer-to-Peer Enterprises (dissertation) – "Forget the Data Scientist, I need a Data Janitor." (Big Data Borat)

If we look back at the 2011–2013 upheavals we see bursts of "social media" activity. From Tahir to Taksim, from Tel-Aviv to Madrid, from Sofia to São Paulo, what they have in common is communication

---

5   See Trebor Scholz, "The Rise of Platform Cooperativism," in *Uberworked and Underpaid: How Workers are Disrupting the Digital Economy* (Cambridge: Polity, 2017), 155–92 and the related event, Platform Cooperativism: The Internet, Ownership, Democracy, The New School, New York, November 13–14, 2015, HTTP://PLATFORMCOOP.NET. See also, *freethought* – a collective formed in 2011 by Irit Rogoff, Stefano Harney, Adrian Heathfield, Massimiliano Mollona, Louis Moreno, and Nora Sternfeld, HTTP://FREE-THOUGHT-COLLECTIVE.ORG.

peaks, which fade away soon after the initial excitement, much in line with the festival economy that drives the Society of the Event. Corporate social networking platforms such as Twitter and Facebook are considered useful to spread rumors, forward pictures, file reports, and comment on established media (including the Web). But no matter how intense the street events may have been, they often do not go beyond "short ties." As temporary autonomous spaces they feel like carnivalesque ruptures of everyday life and are perhaps best understood as revolts without consequences.

In the aftermath of 2011 we've seen a growing discontent with event-centered movements. The question of how to reach a critical mass that goes beyond the celebration of temporary euphoria is essential here. How can we get over the obvious statements about the weather and other meta fluctuations (from Zeitgeist to astrology)? Instead of contrasting the Leninist party model with the anarcho-horizontalist celebration of the general assembly, we propose to integrate the general network intellect into the organization debate. We've moved on a good 150 years since the Marx-Bakunin debates.

It is time to integrate technology into the social tissue and no longer reduce computers and smart phones to broadcasting devices. As so many know, either tacitly or explicitly, technologies are agents of change. To understand social transformation therefore requires an understanding of technology. Harold Innis and Marshall McLuhan both knew this well. It is therefore not unreasonable to say that media theory provides a reservoir of diagnostic concepts and methods to assist those making interventions against regimes of control and exploitation. We would even go one step further: don't just rehash concepts on file, but invent your own by deducing the correspondence between concepts and problems as they manifest within your own media universe of expression. Find sites of conflict, passion, and tension, and you'll soon get a rush of thought to the brain.

The organized networks model that we propose in this book is first and foremost a communication tool to get things done. We are aware that this proposal runs into trouble when (tens of) thousands of users start getting involved. Once you hit that kind of scale the Event takes over. The "orgnets" concept (short for organized networks) is clear and simple: instead of further exploiting the weak ties inside the dominant social networking sites, orgnets emphasize intensive collaborations within a limited group of engaged users with the aim of getting things done. The internet's potential should not be limited to corporate platforms that are out to resell our private data in exchange for free use. That option gives you silos ripe for NSA raids. Orgnets are neither avant-garde nor inward-looking cells. What's emphasized is the word "organ." With this we do not mean a New Age-gesture of a return to nature or a regression into the (societal) body. Neither is it a reference to Aristotle's six volume work called the *Organon*. Even less does it refer to the tired notion of the "body without organs" (or Žižek's reversal, for that matter). The organ of orgnets is a social-technical device through which projects are developed, relations built, and interventions made. Here, we are speaking of the conjunction between software cultures and social desires. Crucial to this relation is the question of algorithmic architectures, something largely overlooked by many activist movements who adopt – in what seems a carefree manner – commercially motivated and politically compromised social media software such as Facebook, Twitter, and Google+.

Today's revolts no longer result from extensive organizational preparations in the background, neither do they produce new networks of "long ties." They do, however, often emerge from a collective unconscious of accumulated discontent. The informal networks that unzip the tweets and create events are the real forces behind the growing list of "global uprisings," from M15 in Spain, Gezi Park in Istanbul, to "yellow umbrellas" in Hong Kong. Think of the public protests in São Paulo: initially a response to an increase in the costs of public transport, the underlying motivation behind such demonstrations was a longstanding malaise stemming from social inequalities and economic privileges bestowed upon a corrupt elite. What's left is a shared feeling: the birth of yet another generation, though one not limited to age or even necessarily class or political persuasions. Even though small groups have often worked on the issues for many years, their efforts are usually focused on advocacy work, designing campaigns, doing traditional media work, or attending to those who are immediately affected by the crisis on the ground. Important work, but not precisely about preparing for the Big Riot.

Is it wishing for too much to want sustainable forms of organization when the world seems to be in perpetual flux, if not on the brink of total chaos? Very little stability defines labor and life as we know it. Ideologies have been on the run for decades. So too are political networks amongst activists. At best we can speak of a blossoming of unexpected temporary coalitions. What we need to focus on in the years to come is time-in-between, the long intervals when there is time to build sustainable networks, exchange ideas, set up working groups, and realize the impossible, on the spot. How might such a long-term strategy be conceived and orchestrated within the logic of networks?

We can complain about social media causing loneliness, but without a thorough re-examination of social media architectures such sociological observations can easily turn into forms of resentment. What presents itself as social media critique these days often leaves users with a feeling of guilt, with nowhere to go, except to return to the same old "friends" on Facebook or "followers" on Twitter. As much as mainstream social media platforms come with an almost guaranteed capacity to scale as mass networking devices, they are not without serious problems that many are now familiar with: security of communication (infiltration, surveillance, and a willful disregard of privacy), logic or structure of communication (micro-chatting among friends coupled with broadcasting notices for the many subscribed to the cloud), and an economy of "free labor" (user generated data, or "the social production of value").[6]

While there has been some blossoming of social media alternatives such as Lorea (www.lorea.org), which is widely used among activists in Spain, other efforts such as Diaspora ended quite disastrously after successfully raising $200,641 in development funds through Kickstarter but failing to gain widespread traction among activists, until an overall implosion of the project after one of its founders committed suicide. The increasing migration of youngsters to Instagram (a subsidiary of Facebook) and Snapchat was probably inevitable (irrespective of whether the NSA leak happened or not). But as April Glaser and Libby Reinish note in a *Slate* column, these social media alternatives "all use centralized servers that are incredibly easy to spy on."[7]

Current social media architectures have a tendency to incite passive-aggressive behavior. Users monitor, at a safe distance, what others

---

6   Tiziana Terranova, "Free Labor: Producing Culture for the Digital Economy," *Social Text* 18.2 (2000): 33–58

7   April Glaser and Libby Reinish, "How to Block the NSA from your Friends List," *Slate*, June 17, 2013, http://www.slate.com/blogs/future_tense/2013/06/17/identi_ca_diaspora_and_friendica_are_more_secure_alternatives_to_facebook.html.

are doing while constantly fine-tuning their envy levels. All we're able to do easily is to update our profile and tell the world what we're up to. In this "sharing" culture our virtual empathy is on display, but not a lot else. "She really ain't all that. Why does all the great stuff happen to her and not me?" Organized networks radically break with the updating and monitoring logic and shift attention away from watching and following diffuse networks to getting things done, together. There is more in this world than self-improvement and empowerment. Network architectures need to move away from the user-centered approach and instead develop a task-related design undertaken in protected mode.

Three months into the Edward Snowden/NSA scandal Slavoj Žižek wrote in *The Guardian* "we need a new international network to organise the protection of whistleblowers and the dissemination of their message." Note that the two central concepts of our argument are utilized here: a network that organizes. Once we have all agreed on this task it is important to push the discussion further and zoom in on the organizational dimension of this timely effort. It can be an easy rhetorical move to emphasize what has already been tried, but we nonetheless need to do that.

One of the first observations we need to make is how Anonymous is the missing element in Žižek's list of Assange, Manning, and Snowden. Despite several setbacks – including more recent associations with the Alt-right movement – Anonymous remains an effective distributed effort to uncover secrets and publicize them, breaking with the neoliberal assumption of the individual as hero who operates out of a subjective impulse to crack the code in order to make sensitive material public.[8] The big advance of anonymous networks is that they depart from the old school logic of print and broadcasting media that needs to personalize their stories, thereby creating one celebrity after the other. Anonymous is many, not just Lulzsec.

We also need to look into the many (failed) clones of WikiLeaks and how specific ones, such as BalkanLeaks, manage to survive. There is also GlobaLeaks and the outstanding technical debate about how to build functioning anonymous submission gateways. It has been

8   See the Nettime mailing list thread on "The alt-righty and the death of counterculture," July 2017, HTTPS://NETTIME.ORG/LISTS-ARCHIVES/NET-TIME-L-1707/THREADS.HTML.

widely noted that WikiLeaks itself is a disastrous model because of the personality cult of its founder and editor-in-chief, Julian Assange, whose track record of failed collaborations and fallouts is impressive. Apart from this "governance" debate, we need to look further into the question of what the "network" model, in this context, precisely entails. A step that WikiLeaks never dared to take is the one of national branches, based either in nation-states or linguistic territories.

To run a virtual global advocacy network, as Žižek suggests, looks sexy because of its cost-effective, flexible nature. But the small scale of these Single Person Organizations (SPOs) also makes it hard to lobby in various directions and create new coalitions. Existing networks of national digital civil rights organizations should play a role here, yet haven't so far. And it is important to discuss first why the US-organization Electronic Frontier Foundation, the European Digital Rights network, or the Chaos Computer Club for that matter have not yet created an appealing campaign that makes it possible for artists, intellectuals, writers, journalists, designers, hackers, and other irregulars to coordinate efforts, despite their differences. The same can be said of Transparency International and journalist trade unions. The IT nature of the proponents seems to make it hard for existing bodies to take up the task to protect this new form of activism.

Design your Power

7,136,376 people like this. Sign up to see what your friends like.
  – Facebook

I want to return to a world without recommendation algorithms.
  – Jenny Schaffer (VICE)

There is a scenario that can influence the work and lives of billions. It is a simple reversal of the dominant social media logic of monopolies such as Facebook, Twitter, and Google. Instead of growing networks through "weak ties" users concentrate their efforts on small groups in order to get things done: a collective move from communication to social action, from weak ties to strong links. So far network gurus have only looked to the ever-growing imaginary of connection. Software and algorithms are designed to expand our engagement with the "link economy," though without any form of remuneration that arises from the capture of data and extraction of value.

But what's the use of endlessly maintaining the network of 500+ "friends," where your primary occupation becomes "working for the timeline"? For all the pictures we upload and status updates we generate, our primary signal to friends we've never met is that we're still in the rat race: look at me, I am still alive, do not forget me. Tragically, the cultivation of the celebrity-self is even more forgettable than the unobtainable juice of fame we secretly slather on to our increasingly numb membrane of desire.

We should start sabotaging the pressure to update and grow our networks. Strategies, if not devices, are required that short-cut the implicit competition that so often compels us to act. The proposal here it to intensify what's already there and collaborate – instead of merely communicate – in ways that ensure existence is a force to be reckoned with. Call it a lingering passion to invent. The concept of organized networks is first and foremost an Unidentified Theoretical Object (Adilkno), a space of potentialities that can be opened – and closed again. Read it as a proposal to undermine the widely-felt Fear of Missing Out.

Amalgamating the words "organization" and "network," the concept of orgnets is something we developed in 2005 as a response to the rise of the "social networking" paradigm and orthodox ideas in management circles about the "networked organization." The term can be read as a variation and upgrade of the popularity and mystique that surrounds "organized crime," while intersecting with the more imaginative but slightly conceptual term "organized innocence" (as described by the Adilkno collective in their book *Media Archive* from 1998). Needless to say, orgnets are both virtual and real. They are as much living data, crunching away on hard-disks, as they are hardcore urban tribes, non-identities, invisible for non-members.

Orgnets have grown in response to European offline romanticism and assembly strategies from Occupy activists. Meeting in-real-life ("breast-to-breast") is touching yet expensive and often impossible to arrange on the hop. Most collaborations these days, if serious, are not touristic in nature anyway. Leave those junkets for the coterie clinging to the vestige of power bestowed upon boardrooms. There is a tragic, harsh element in the fact that more often than not we don't coincide in the same room, building, city, or continent. This is the rotten reality of our global existence.

Organized networks are out there. They exist. But they should still be read as a proposal. This is why we emphasize the design element. Please come on board to collectively define what orgnets could be all about. The concept is an open invitation to rethink how we structure our social lives mediated through technical infrastructures.

Whereas it is possible to interpret the rich history of humankind as orgnets, from clans and villages to secret societies, collectives, and smart mobs, we prefer to emphasize the 21st century blend of technology and the social. Orgnets have appeared on the scene in a time of high uncertainty. Not only do we have the catastrophe of planetary life driving fear into the soul of the future. But we also have what seems a broader social incapacity to act. And this is partly a result of the problem of traditional institutional forms grappling with the challenge – still – of a world that is deeply networked by digital media.

Witness, for example, the crisis of conventional organizations such as the trade union, the political party, the church, and the social movement. Losing credibility by the day, increasingly decoupled from their constituencies, and no longer able to galvanize collective passion to mobilize action. The primary pillars of social organization that defined the 19th and 20th centuries have struggled to reinvent themselves to address the complexities that define our times. This is where orgnets step in.

Networks are not goals in themselves and are made subordinate to the organizational purpose. Internet and smart-phone based communication was once new and exciting. This caused some distraction, but the widespread enthusiasm they once elicited is definitely now on the wane. Distraction itself is becoming boring. The positive side of networks – in comparison to the group – remains its open, informal architecture. However, what networks need to "learn" is how to split-off or "fork" once they start getting too big. Scale can become the enemy. At this point networks typically enter the danger-zone of losing focus. Intelligent software can assist us to dissolve connections, close conversations, and delete groups once their task is over. We should never be afraid to end the party.

2.

THE LEGACY OF
TACTICAL MEDIA

The immanent logic of social development points to a
totally technicized life as its final stage.
— Max Horkheimer, *Dawn & Decline*, 1978.

The problem of revolutionary organization is the
problem of setting up an institutional machine whose
distinctive features would be a theory and practice that
ensured its not having to depend on the various social
structures ....
— Félix Guattari, *Psychoanalysis and Transversality*, 1972.

This is the current form of waiting. A tedious interval
without patience or hope.
— Vilém Flusser, *Post-History*, 1983.

Tactical media will never become organized. The World Federation of Tactical Media was not a missed opportunity; luckily we never even got close. Yet tactical media subsists in an era of logistical media of orientation and control. Logistical media organize the world in the image of capital accumulation made operational in real-time. Tactical media, on the other hand, will never organize as new institutional forms. Tactical media, that post-Berlin Wall child of multi-media and internet practiced by activists, designers and artists, hackers, and video enthusiasts, refused to make history. This does not mean that institutional forms will never become tactical. Herein lies the legacy of tactical media as a strategic intervention. The practice of tactics from within the horizon of financial capitalism, for example, generates the possibility of alternative models of distribution that seed the production of the commons.[1] Becoming tactical is the game of appearing as a singular entity – and disappearing at the moment of over-exposure.

Tactical media offer a model for collective investigation on transnational scales. Indymedia was emblematic of a networked model of organization galvanized by the anti-WTO protests in 1999.[2] With grass-root nodes distributed in cities across the world, Indymedia became renowned for participatory forms of critical journalism that offered an alternative media perspective on anti- and counter-globalization movements. Its extensive commentary and internal debates on global justice issues on racism and sexism connected with local conditions for political activists. With its mixture of subdomains on a central server with other sites and nodes on nationally hosted servers, Indymedia was a proto-form of bringing global computer networks together with temporary Independent Media Centers in places of (global) resistance.

A decade later the gathering of protesting masses on the streets turned virtual. WikiLeaks, for example, shifted the emphasis from independent coverage of protests to investigative journalism based on publishing anonymously supplied documents revealing the inner workings of global elites, finance, and military apparatuses: a nomadic enterprise without any office, based on free software models of collaborative

---

1   The example here is the Robin Hood Minor Asset Hedge Fund, HTTP://www.
ROBINHOODCOOP.ORG/.

2   HTTPS://EN.WIKIPEDIA.ORG/WIKI/INDEPENDENT_MEDIA_CENTER.

production.[3] Other relevant examples include Anonymous, the concentration of the energy of the crowds into D-DOS attacks, online versions of direct action, and early crypto-currency experiments with Bitcoin. What these diverse hacktivist initiatives have in common is their implicit critique of the legally incorporated NGO model of local, regional, national offices, often coordinated out of a global HQ (such as Amnesty, Greenpeace, Wikimedia).

Tactical media remains exemplary as a collective practice of identifying and exploiting vulnerabilities in targets it seeks to oppose. Its refusal to scale up can be read as a weakness, but the beauty of the 1970s "small is beautiful" discourse made it possible to mobilize interventions by the few that could then scale up and resonate at global levels since it operated on the terrain of the symbolic. This kind of simulation of protest runs into fatigue when it is not followed up with substantive structural change in people's lived conditions. In this regard tactical media can be faulted for a kind of lack of collective imagination to organize in ways beyond the immediacy of the event.

It is not that tactical media didn't have its "Lean & Agile" methods. It did. Take a look at the diagram on the inside cover of *Handbuch der Kommunikations Guerilla*, a tactical media handbook in which "Baader-Meinhof Meets Baudrillard."[4] We read the following: "Some of the many tactical methods and principles catalogued here include sniping, cross dressing, happenings, collage and camouflage as techniques of subversive affirmation. Everyone must become a clown, but these willful acts of playfulness could press into middle age." The question of renewing energy and strategy within an era of post-media fatigue is at the core of our inquiry in this chapter.

Duration of Infrastructure vs. Singularity of the Event

Over the years tactical media lost most of its autonomous infrastructure. Once squatted spaces had to be rented back. Computer servers

---

3   See Jennifer Cloer, "Author Gabriella Coleman Expands on Role of Linux in Hacker Culture," Linux.com, December 17, 2012, HTTPS://WWW.LINUX.COM/ NEWS/FEATURED-BLOGS/185-JENNIFER-CLOER/682035-AUTHOR-GABRIELLA-COLEMAN-EXPANDS-ON-ROLE-OF-LINUX-IN-HACKER-CULTURE.

4   autonomous a.f.r.i.k.a. gruppe (aka Luther Blissett and Sonja Brünzels), *Handbuch der Kommunikations Guerilla* (Berlin: Verlag Libertäre Assoziation, 1997).

broke down and were not replaced, websites disappeared, as did video editing facilities and free radio studios. If tactical media were a form of intervention, this was often only at the level of the sign or the symbolic. In that sense it was locked into a post-war concept of the media as the territory of ideology and institutional condition of massive expansion. The error of tactical media was to overlook and ignore media of orientation, which in other circumstances would have consisted of an infrastructural solution.

In the new festival economy infrastructures are created in an instant pop-up manner. The Burning Man festival embodies this model of distributed management to the extreme. Where is infrastructure after the orgy? Tactical media knew very well how to organize logistically, but not strategically to sustain an alternative media system. Temporary infrastructure is a ludicrous symbol of complete waste. Why build up a protest for 200,000 people only to see it disappear tomorrow? Just because you can? To publicly display your ability to vanish without consequence? Because we're afraid of becoming bored? Unable to hang around for the long fight? Factors such as these suggest that other infrastructural, institutional, and economic conditions were prevailing in the background.

Tactical media offered a temporary departure from these routines, caught up in the seduction of the event. The event is an intensive collective mobilization of imagination. The strategic choices made here will influence the political culture for decades to come: are social movements in need of more solid infrastructure (buildings, data centers, education facilities, and other commons of sorts), or will counter-power be built-up in the staging of ever bigger and significant events?

The event as spectacle is no doubt related to the rise of the "creative city" – the latter stretches the temporality of the event in the short-life form of the start-up and incubator. There is an infrastructure of the event, but nothing that remains as a legacy architecture. Curiously, at its peak in the late nineties tactical media pointed to politics as that which operates at the level of the sign. Nowadays, the sign is well and truly depleted of authority to command allegiance to either identity or ideas. Politics is territorial, material, and predominantly about occupation. This takes place in a context of cities driven by real estate bubbles, gentrification, massive cuts in cultural funding, and the subtraction of space for public activities. The city has become a zone of exclusion where permission is required in advance of the act. This does not bode well for tactical media as we knew it.

Tactical Media and the Problem of Disorganization

Tactical media produced the network as an information and social architecture of exchange. But the preferred organizational form of this network was again the event, often in the shape of the hit-and-run media action. Tactical media produced the meme, which remains a powerful internet based mode of dissemination and multiplication. The form of the meme has also been absorbed into the political economy of advertising and social media marketing. In so doing, it became anemic as a political device. This in itself is a curious residue because it points to the continuity of a form and of a practice that suggests that content still counts. But when was the last time you encountered a radical meme that disrupted perception on a mass scale? For a second we thought that Bitcoin creator Satoshi Nakamoto was revealed as Craig Steven Wright – an angry and arrogant crypto expert and IT consultant buried away in a Sydney suburb.[5] But this turns out to be yet another hoax in the long trail of media stories in pursuit of boredom revisited. So much for rattling the threshold of perception.

How to sustain an idea over time? Occupy was a meme, but not a network that could endure in ways that embody the critique of global finance capital. Yet global intellectual celebrities such as Thomas Piketty and Yanis Varoufakis were able to do so in ways similar to Al Gore or Naomi Klein's stand against global warming. The event-based temporality is already a given in the very idea of occupying. The act of occupation should be consolidated by a take-over or admission of failure followed by dissolvement. In the 1990s tactical media was understood as a form of spontaneous blossoming. Tactical media knew how to mobilize resources. Finance, hardware, and event infrastructure were the staples of hackfests and the like. This was helped along

---

5   See Sarah Jeong, "Satoshi's PGP Keys are Probably Backdated and Point to a Hoax," *Motherboard*, December 9, 2015, HTTP://MOTHERBOARD.VICE.COM/ READ/SATOSHIS-PGP-KEYS-ARE-PROBABLY-BACKDATED-AND-POINT-TO-A-HOAX. See also Andrew O'Hagan, "The Satoshi Affair," *London Review of Books* 38.13 (June 2016): 7–28, HTTPS://WWW.LRB.CO.UK/V38/N13/ANDREW-OHAGAN/ THE-SATOSHI-AFFAIR.

by an enormous implosion of costs related to media production and the ubiquity of tech-driven consumption related to the camcorder revolution and increasing miniaturization.[6] The age of mechanical reproduction was prohibitive in terms of costs and equipment. The rise of the PC enabled everyone to become a producer. This was the democratization of media realized as practice.

Tactical media has organized the archive but not itself.[7] There was no mechanism or repertoire of practices that enabled collective self-organization over time and space. Snowden embodies the tactical media instance to the supreme end: a massive hack instigated by the individual who is then left stranded and dependent on the benevolence of accommodating states.[8] There is no distribution of responsibility in such a model. Anonymous is the counter example to this. Anonymous has cells that occasionally connect in the cloud. But is there a form or practice between these two extreme polls of the unexpected act of the individual and the enormously diffuse action of anonymity? Can organization have a face?

That tactical media could build temporary coalitions tells us that there can be a protocological interoperability not limited to the fantasies of SAP, Oracle, and the world of enterprise software. The major

---

6   See Rey Chow, "Listening Otherwise, Music Miniaturized: A Different Type of Question about Revolution," in *Writing Diaspora: Tactics of Intervention in Contemporary Cultural* Studies (University of Indiana Press, 1993), 144–65.

7   This archiving happens, interestingly enough, when the term "tactical media" is on the cusp of being forgotten 25 years after it emerged. To historicize now holds significance and importance for the artists, activists, and scholarly community that seeks to create continuities that signal relations over fragmentation. Such historicizing also parallels trends in the retro-mania of our times that revives previous cultural epochs, whether this is food, fashion, architecture, music, or drugs. We can meet each other again at the nineties party. At stake for tactical media, at least, is the question of politics and technology. Whether historicizing this period can happen without sacrificing the collective cultivation of political intervention through technological means remains to be seen.

8   See Florian Sprenger, *The Politics of Micro-Decisions: Edward Snowden, Net Neutrality and the Architectures of the Internet*, trans. Valentine A. Pakis (Lüneburg: Meson Press, 2015), HTTP://MESON.PRESS/BOOKS/THE-POLITICS-OF-MICRO-DECISIONS/.

software companies and developers of enterprise systems have ruth-lessly exploited this. But tactical media shied away from building an alternative communications infrastructure in ways that scaled. There is not a single tactical media success story in this regard. Tactical me-dia abandoned its creations as rapidly as it built them. Tactical media explored the user status and then handed over the keys.

There was no long march through the institutions, even if tacti-cal media was often a response to the command and control logic of organization. Some generations would rather disappear. Building on already existing infrastructure there was always a parasitical element to tactical media. But there is a political safety net that comes with the parasite who lives off its host. Despite its fine eye for the interven-tion in the time of the present, tactical media was not able to address or clearly formulate what was at stake. This requires both a sense of historical conditions but also an imaginative projection into the fu-ture with regard to transformations that would manifest as successive economic crises. The tech-wreck was just around the corner, 9/11 pro-duced a digital society of surveillance, and the advance of neoliberal economic agendas gutted support for cultural work.

The main aim of tactical media was to share aesthetic content made possible by easy access to media production and communication sys-tems. This is another key legacy of tactical media. The economy of sharing is indeed abundant whether it is Airbnb, Uber, or the "shar-ing" platforms. In this respect an infrastructural continuity certainly exists, but it is not one owned by the world of tactical media and not especially enamored with the act of hacks.

Can we be asking too much of tactical media? It was an activist strategy in a period of destruction and decline and was remarkable for what it achieved. But it was always going to be low-scale with a relatively small and geographically dispersed number of participants. The fact that an anthology of essays on tactical media is published by MIT Press suggests an element of organization – of cultural memo-ry and also myth about what tactical was and could still be today.[9] Journal articles are published, university courses designed, and prac-tices reproduced within the cultural sector as a post avant-garde art movement. But this is organization of the formal sector and not the

---

9   Erik Kluitenberg and David Garcia (eds), *Tactical Media Anthology* (Cambridge, Mass.: MIT Press, forthcoming).

activist scene *per se*. To this end the host ultimately benefitted from the parasite. This was consciously part of the design. Tactical media was never re-appropriated or institutionalized, though its lessons were absorbed and integrated, for example, in the post-internet art movement. Apart from this occasional instances, tactical media had long disappeared and passed through its Winter years. There was a failed link to the dotcom start-up world and a conscious decision to not play into it. Individuals were participating here and there as programmers and designers, but soon lost their jobs. By 2001 it was all over.

The unconscious of tactical media consists of a strategy of disappearance, but this is not one that increasingly defines the corporate landscape of the IT sector. The stack hums along in the background. Disappearance in this model is about thorough integration into daily routines whose economy is driven by the social reproduction of value captured by technical systems. As Bruce Sterling writes, "Google and Facebook don't have 'users' or 'customers.' Instead, they have participants under machine surveillance, 'prosumers' whose activities are algorithmically combined with Big Data silos."[10] Despite all critique that says otherwise, Facebook and Twitter remain today the primary media of tactical intervention and political mobilization. The networked 2.0 platform of Facebook organizes activism into "groups," which function to contain the distributed logic of the network. The world is perceived through the network, but the network is circumscribed and so too is the intervention. Rarely does it infect or impregnate the social body in the way that Facebook's invisible integration does.

The Laws of Organization

The question of organization is also a question of occupying time. There is a long durée assumed in the work called organization. There is a plan, a shape, a vision. There are targets, goals, benchmarks, and, above all, maintenance issues. These are often all highly unattractive terms and practices for media activists. Lean and Agile organizations are Enterprises. They have a mission statement and operational logic tailored to the service economy. "Agility has emerged as the successor to mass production."[11] Did tactical media fail because it didn't have a

---

10   Bruce Sterling, *The Epic Struggle of the Internet of Things* (Moscow: Strelka Press, 2014), 10.

11   ASQ, "Agile Enterprise versus Lean Enterprise Tutorial," HTTP://ASQ.ORG/

positive reward system in place? It neglected its "continuous improvement reviews." Tactical media was never really a model, which is one mechanism that tests organization: the model either fails or works. Tactical media could therefore never fail, but it did. So is there retrospectively a model to be exhumed from the tactical media moment? Tool kits were produced in relative abundance, and the multidisciplinary variation of this is still around. The integration of politics, aesthetics, and technology, which for a long time were seen as distinct spheres, are one of the key legacies of tactical media. These are now skills that to an extent are integrated into the global labor force, with the exception of politics which requires organization. The Lean & Agile Enterprise is a global model, but it is not one whose flexibility is amendable to the organization of politics, which is seen to be disruptive and synonymous with disorganization.

Tactical media was fantastic at setting loose playful propositions for political and social alternatives. In this regard it was a factory of blueprints from cyber-feminism to free software and the production of the commons. The legacy of tactical media was that other organizations and entities went on to realize these blueprints. Tactical media subsists within this legacy, yet without responsibility in the sense of attribution. Aside from exceptions such as The Yes Men, it is for the most part a movement without monuments. The auto-deconstructivist impulse makes it impossible to build narratives of continuity that provide a basis upon which organizations are founded. This habit is one that may be generationally specific to those who came of age in post-Cold War period, and is perhaps not one that resonates for or connects with previous or successive generations. In this regard, the strategy for post-tactical media requires an exit from self-affirmation.

In the context of a closed media world, which goes against the tinker culture of tactical media, the question of organization is even more pressing. Where once tactical media could produce alternative media of communication, nowadays the revolution will be among Friends. This is where the prospect of organization through distributed infrastructures such as the blockchain technology become important to consider. The network can fork, it can multiply itself, and it can reconfigure new parameters. The blockchain as defined by the Bitcoin community has the capacity to process data securely, in a decentralized

---

LEARN-ABOUT-QUALITY/LEAN/OVERVIEW/AGILE-VS-LEAN-TUTORIAL.HTML.

fashion, and rapidly multiply. If organization is always immanent to communication technology, this presents an interesting prospect, despite all the right-wing neolibertarian ideologies that surround the crypto-currency hype. There is another blockchain, as the MoneyLab network, coordinated by the Institute of Network Cultures and many of its affiliated initiatives, are trying to prove.[12]

Tactical media could have organized itself as a forum with memberships and the capacity to decide on expulsions. It is romantic to suppose that something coherent like this was possible. But from the start it was already a network and this was a key struggle because its form was not immediately or obviously associated with the strategic question of organization and more coordinated practices of governance that must at some point commit to the decision. The flat architecture of forums are often running in parallel, not indicating any hierarchy in terms of importance. Ranking sites such as Slashdot, Belong and Hacker News (CloudFlare) or its commercial version, TheRanking.com, function through the recommendation economy. What these sites lack is any aesthetic sensibility – they are never going to be sites that excite the desire for visual stimulation. These are old school content aggregators. Algorithmic automation is not a substitute for political decision. There is a critical legitimacy to be had from the analysis of algorithmic architectures, but the brilliance to be found in some of this work often assumes the black box has something to say about the operational complexity of the world. Unfortunately this is rarely if ever the case.

Paradoxically, in this time of ubiquitous social media and its culture of sharing what, at the end of day, is shared? Why isn't there organization? Sharing without consequences will not produce the organization of networks. The current fad in recent years that defines the lulz culture of cat memes may be high on camp irony, but this is a sort of nihilistic irony, again without political consequence. And this, coming from the often politically righteous proponents of Anonymous – a social-culture movement that shuns a commitment to seriousness. Everything is a joke. But this doesn't help with the problem of organization.

What kind of organization can work? There's Amazon's Mechanical Turk, which gets the job of low-skill data entry jobs done to the point

---

12   MoneyLab, Institute of Network Cultures, HTTP://NETWORKCULTURES.ORG/ MONEYLAB/.

at which the human is indistinguishable from the machine. But this is hardly a model for organizing politics. It is one of many that demonstrates how the economy can be organized with network architectures. But the sociality of dispute and political disagreement does not lend itself easily to the interface defaults of networks. There is a division of labor that accompanies the development of a well-designed task, but why is this antithetical for network politics? The mytheme of horizontality didn't do organized politics any favors. How does a cell know that it's part of a larger entity? This is less about the friendly sharing of kooky cats, couches, or car rides than it is about a shared political vision. For all the depth over decades and centuries of developing different strains of political theory and practice, it's remarkable the extent to which such modes of critical consciousness have not been able to transpose over and translate into networked infrastructures of communication, economy, and life. The internet, broadly put, remains a barrier to implementation. Yet it has clearly transformed the organization of this world. Are the multitudes really so incapable of following through on a plan? Do they really want another possible world that overcomes borders, destruction, and distraction? People give up so soon. The dissipation machine spreads energies, desire, and attention into clouds of dispersed confusion and impressions. The violence of commitment is what fails within the computational universe that, oddly, is one in which recursive feedback is made inoperable. There is no intentionality.

Belonging alone will not be sufficient to the work of organization. Membership offers no guarantees. Tactical media wasn't about citizen empowerment, even if at times it looked that way. It was much more basic – it was about doing things together. The smart city excites the neoliberal subject to measure the weather, but really the bots and sensors do that on their own. The key to organization is about creating new forms of sociality not reducible to measurable units. Beyond automation is the almost erotic gathering of bodies into social assemblies and the like.

Perhaps there is an ontology of power that is not amenable to distributed systems and technics of communication. The propensity or disposition of power is one often associated with the tyrant or coercive group. Moribund, corrupt, yet always able to decide, mobilize, and importantly implement. Networks, by contrast, such as Occupy dissolve or distribute power and this is contrary to the idea that the networks should become more powerful and visible themselves. The ethos of collective decision making through distributed communication systems, while also gaining power, is wishful thinking. The ideology of leaderlessness is a curious paradox: at once unobtainable as an end goal, but nonetheless suggestive of a structure in place to produce ideology indicating a hierarchical set of relations.[13] If ideology is produced otherwise, which is to say not through hierarchy but through horizontality, why then does power elude the organization of networks?

The Future of Autonomy after Refusal

If you are condemned to the café or subject to the creative alienation of the co-working space and the tyranny of transparency the open plan office offers, if you are left alone in the library or the disillusioned intern in what used to be an energetic architecture office that is now strung out on the upkeep of real estate rents, then where to go for autonomous organization? The tendency to organize within the walled gardens of social media isn't going to cut it. No restoration of the social going on there. But refusal is also not such an option when that means casting yourself adrift as an individual scurrying below the radar of the NSA surveillance machine. More than ever, it seems that autonomous organization eludes collective consciousness let alone individual desires for independence. In part this is a question prevailing economic conditions. But even more than this is the problem of infrastructure.

Previously one could set up pirate radio, social centers, hacker spaces (such as the once infamous ASCII in Amsterdam), hang out on the fringes of festivals, make happenings in the underground cafes, build art spaces and studios, maintain theatre and performance spaces, run off zines on your own printing press or sever. These were all infrastructure of autonomy for the expression of radical cultures of one kind or another. Nowadays, that universe hardly registers on the horizon of hipsters and their struggles. The real estate logic savaged pretty much all of these initiatives. The idea of searching out new urban spaces as gentrification gobbled up one area after another is no longer viable. Gentrification has spread across the entire urban

---

13   On the question of leadership for movements, see Michael Hardt and Antonio Negri, *Assembly* (Oxford and New York: Oxford University Press, 2017).

space, no longer making possible the movement to other down and out suburbs to set up your studio or practice all over again. Not only is this the case within cities, but increasingly cities across the world no matter where you are have become subsumed by the financialization of urban space. This can only be negated through a severe financial crisis. But what is one supposed to do? Hold your breath for a decade waiting for that to happen? To project hope on mass poverty and mass alienation is an extreme form of negation that is certain to deplete the life of the soul. So what other options are there? This is the problematic of the current conjuncture that requires tactical, strategic, and perhaps even logistical moves beyond the submission of refusal.

The use of web-based interfaces for organization is really not an option for tactical media these days. Facebook will never get you there. That's not say that the digital won't play a role. One only need recall the use of mobile phones and distributed Bluetooth connections using FireChat in the pro-democracy movements in the Hong Kong summer of 2014. But once again, political momentum was not able to easily sustain itself beyond a few months. Let us be clear, we are not pinning this to some kind of techno-solutionism. But there's no doubt that tactical media, after all, require media. The Silicon Valley tactic is to move into the post-media universe of the Internet of Things, colonizing agricultural industries, urban transport systems, logistical supply chains, educational settings, healthcare services, the quantified self movement. On it goes. This is in part the desperation of capital accumulation as it seeks to extract value from any and every thing. So if Silicon Valley are on to this disruption tactic that raids the last vestiges of the common, where to go for a radical media? Is this an occasion to devise a non-disruptive agenda for tactical media beyond appropriation and cooptation? To be drawn into some kind of reformist agenda is also undesirable here.

At this point, we enlist our concept of organized networks as an example of tactical media recast into the present. Offsetting the preoccupation in much of net-culture in the nineties with tactical media as the vanguard of experimental intervention, we emphasize the strategic dimension organization as a way to ask how new institutional forms are invented within distributed network environments. There is a tendency for tactical media to vanish after the event. Now we find ourselves within a global context of protracted financial crisis, ecological disaster, and a situation in which social and economic despair is a norm, no longer exclusive to the down-trodden and impoverished members of this world. Organization and media are the core components of infrastructure of consequence for those without means. This was for a while the delirious promise of Web 2.0, but that model of participatory culture and user-generated content was suited for the tech-sector vision of a neoliberal individual who did it all for free, who in the end needs to be serviced by the R&D machine of Silicon Valley. This business again suits the commercial interests of some, but it doesn't do a whole lot for the collective practice of autonomy.

The contemporary arts version of this narrative, as neatly summarized by Claire Bishop, did its best to cultivate a cozy account of the citizen as a participant in the social spectacle of the museum and gallery space. But again, we see no conceptual or technical resources here for the collective coordination of infrastructure that unsettles the status quo while remaining robust enough to withstand serious setbacks and eventual collapse. This means building an infrastructure where no individual can run away with the keys. We are in a prolonged moment in which cultural and political organization is trapped in an "empire of retro" (Simon Reynolds) and "the ache of nostalgia" (Mark Fisher) that lures us into remixing as an anesthetic affect. Occasionally disrupted by the event that resonates on a planetary scale, a false sense of consequence permeates across the social imaginary. But all too quickly we return to the insularity of despair (Bifo).

In this next stage on building infrastructures of autonomy, we need to work on how collective imagination can be turned into a sustainable form of organization with commitment. Current developments in Blockchain architectures are carried by a meta-concept of distributed resources without the right to inspection, despite the hype on transparency. We find it fascinating that the technical here is providing a scaffolding for new models of not just cryptocurrencies but also practices of organization. As much as the banking sector is scrambling to enclose Blockchain within existing payment systems, any number of radical and criminal experiments in new distribution and governance models are pushing against this pressure to disable what for now holds out as a technical ontology of the distributed ledger.

The legacy of tactical media has opened up a space of experimentation without finitude. We believe there is an historical resonance of this ambition that can be mined in the present. If we pose the question of organization this is not to narrow down the options, which is a common criticism and a trap. There is no need to propose a reimagined party-political form as a solution in search of a problem, as Jodi Dean and others have done. Instead, we need to further open up the space of experimentation in the spirit of tactical media to find out what works and what doesn't.

We can indeed learn from the production of failure. But as much as this is advocated by the tech-sector or entrepreneurial political classes, this does not at all mean that a tactical media variation of such a culture of organizational and technological experimentation results in a fine-tuning of the financialization of daily life. The tactical media chapter was not first and foremost about a proof of concept for business ventures. We can count surprisingly few millionaires amongst our friends. The precarity of life for many of those heavily engaged in the time of tactical media has not exactly provided a pathway to middle-class existence. Beyond some of the traditional disciplinary and geographic borders that manage the international division of labor and knowledge production, tactical media showed us that concepts cut across seemingly predetermined divisions and intermingle in unexpected ways that catalyze societies of dissent. Organization coupled with media remain a potent mix beyond submission and control.

3.

DAWN OF THE ORGANIZED NETWORKS

At first glance the concept of "organized networks" appears oxymoronic. In technical terms, all networks are organized. There are founders, administrators, moderators, and active members who all take up roles. Think also back to the early work on cybernetics and the "second order" cybernetics of Bateson and others. Networks consist of modulating relations whose arrangement at any particular time is shaped by the "constitutive outside" of feedback or noise.[1] The order of networks is made up of a continuum of relations governed by interests, passions, affects, and pragmatic necessities of different actors. The network of relations is never static, but this is not to be mistaken for some kind of perpetual fluidity. Ephemerality is

---

1   For elaboration on the concept of the "constitutive outside" as it relates to media theory and the politics of information, see Ned Rossiter, *Organized Networks: Media Theory, Creative Labour, New Institutions* (Rotterdam: NAi Publishers, 2006), 125–31.

not a condition to celebrate for those wishing to function as political agents.

Why should networks get organized? Isn't their chaotic, disorganized nature a good thing that needs to be preserved? Why should the informal atmosphere of a network be disturbed? Don't worry. Organized networks do not yet exist. The concept presented here is to be read as a proposal, a draft, in the process of becoming that needs active steering through disagreement and collective elaboration. What it doesn't require is instant deconstruction. Everyone can do that. Needless to say, organized networks have existed for centuries. Just think of the Jesuits. The history of organized networks can and will be written, but that doesn't advance our inquiry for now. The networks we are talking about here are specific in that they are situated within digital media. They can be characterized by their advanced irrelevance and invisibility for old media and p-in-p (people in power). General network theory might be useful for enlightenment purposes, but it won't answer the issues that new media based social networks face. Does it satisfy you to know that molecules and DNA patterns also network?

There are no networks outside of society. Like all human-techno entities, they are infected by power. Networks are ideal Foucault machines. They undermine power as they produce it. Their diagram of power may operate on a range of scales, traversing intra-local networks and overlapping with transnational insurgencies. No matter how harmless they seem, networks ignite differences. Foucault's dictum: power produces. Translate this over to organized networks and you get the force of invention. Indeed, translation is the condition of invention. Mediology, as defined by Régis Debray, is the practice of invention within the social-technical system of networks.[2] As a collaborative method of immanent critique, mediology assembles a multitude of components upon a network of relations as they coalesce around situated problems and unleashed passions. In this sense, the network constantly escapes attempts of command and control. Such is the entropic variability of networks.

The opposite of organized networks is not chaos. Organized networks routinely intervene into the radical temporality of today's media

<hr />

2   Régis Debray, *Media Manifestos: On the Technological Transmission of Cultural Forms*, trans. Eric Rauth (London and New York: Verso, 1996).

sphere. Short-termism is the prevailing condition that inflicts governments, corporations, and everyday life. Psycho-pharmacology is the bio-technical management of this condition.[3] Organized networks offer another possibility – the possibility of creativity, invention, and purpose not determined in the first instance by the creaking, frequently bewildered grasps at maintaining control, as witnessed across a range of institutions that emerged during the era of the modern state and persist to this day within the complex of the corporate-state, which continues to maintain a monopoly on legitimate violence.

Network users do not see their circle of peers as a sect. Users are not political party members. Quite the opposite. Ties are loose, up to the point of breaking up. Thus the ontology of the user, in so many ways, mirrors the logic of capital. Indeed, the "user" is the identity par excellence of capital that seeks to extract itself from rigid systems of regulation and control. Increasingly the user has become a term that corresponds with the auto-configuration of self-invention. Some would say the user is just a consumer: silent and satisfied, until hell breaks loose. The user is the identity of control by other means. In this respect, the user is the empty vessel awaiting the spectral allure of digital commodity cultures and their promise of "mobility" and "openness." Let us harbor no fantasies: sociality is intimately bound within the dynamic array of technics exerted by the force of capital. Networks are everywhere. The challenge for the foreseeable future is to create new openings, new possibilities, new temporalities, and new spaces within which life may assert its insistence for an ethico-aesthetic existence.

Notworking is Networking

Organized networks should be read as a proposal aimed to replace the problematic term virtual community.[4] Organized networks also supersede the level of individual blogging, whose logic of networks does not correspond with the concept we develop here. It is with some urgency that internal power relations within networks are placed on

---

3   Franco "Bifo" Berardi, "Biopolitics and Connective Mutation," trans. Tiziana Terranova and Melinda Cooper, *Culture Machine* 7 (2005), HTTP://CULTUREMACHINE.TEES.AC.UK/FRM_FI.HTM.

4   See also the introduction and conclusion of Geert Lovink's *My First Recession: Critical Internet Culture in Transition* (Rotterdam: V2_/NAi Publishers, 2003).

the agenda. Only then can we make a clear break with the invisible workings of electronic networks that defined the consensus era. Organized networks are "clouds" of social relationships in which disengagement is pushed to the limit. Community is an idealistic construct and suggests bonding and harmony, which often is simply not there. The same could be said of the post-911 call for "trust."

Networks thrive on diversity and conflict (the notworking), not on unity, and this is what community theorists are unable to reflect on. For community advocates, disagreement equals a disruption of the "constructive" flow of dialogue. It takes effort to reflect on distrust as a productive principle. Indifference between networks is one of the main reasons not to get organized, so this aspect has to be taken seriously. Interaction and involvement are idealistic constructs. What organized networks also do is question the presumed innocence of the chattering and gossiping networks. Networks are not the opposite of organizations in the same way as the real is not opposed to the virtual. Instead, we should analyze networks as an emerging social and cultural form. Networks are "precarious" and this vulnerability should be seen as both its strength and its weakness.

In the information society passivity rules. Browsing, watching, reading, waiting, thinking, deleting, chatting, skipping, and surfing are the default conditions of online life. Total involvement implies madness to the highest degree. What characterizes networks is a shared sense of a potentiality that does not have to be realized. Millions of replies from all to all would cause every network, no matter what architecture, to implode. Within every network there are prolonged periods of interpassivity, interrupted by outbursts of interactivity. Networks foster and reproduce loose relationships – and it's better to face this fact straight in the eye. They are hedonistic machines of promiscuous contacts. Networked multitudes create temporary and voluntary forms of collaboration that transcend but do not necessary disrupt the Age of Disengagement.

The concept of organized networks is useful to enlist for strategic purposes. After a decade of "tactical media" the time has come to scale up the operations of radical media practices. We should all well and truly have emerged from the retro-fantasy of the benevolent welfare state. Networks will never be rewarded and "embedded" in well-funded structures. Just as the modernist avant-garde saw itself punctuating the fringes of society, so have tactical media taken comfort in the idea of targeted micro interventions. Tactical media too often assume to reproduce the curious spatio-temporal dynamic and structural logic of the modern state and industrial capital: difference and renewal from the peripheries. But there's a paradox at work here. Disruptive as their actions may often be, tactical media corroborate the temporal mode of post-Fordist capital: short-termism.

It is retrograde that tactical media in a post-Fordist era continue to operate in terms of ephemerality and the logic of "tactics." Since the punctuated attack model is the dominant condition, tactical media has an affinity with that which it seeks to oppose. This is why tactical media are treated with a kind of benign tolerance. There is a neurotic tendency to disappear. Anything that solidifies is lost in the system. The ideal is to be little more than a temporary glitch, a brief instance of noise or interference. Tactical media set themselves up for exploitation in the same manner that "modders" do in the game industry: both dispense with their knowledge of loop holes in the system for free. They point out the problem, and then run away. Capital is delighted, and thanks the tactical media outfit or nerd-modder for the home improvement.

The paradigm of neoliberalism is extensive throughout the bio-technical apparatus of social life. And this situation is immanent to the operation of radical media cultures, regardless of whether they are willing to admit it. The alarm bells will only start ringing when tactical media cranks up its operations. And when this happens, the organized network emerges as the *modus operandi*. Radical media projects will then escape the bemused paternalism of the state-as-corporation.

But make no mistake, the emergence of organized networks amount to an articulation of info-war. This battle currently revolves around the theme of "sustainability." It is no accident that sustainability is the meme of the moment, since it offers the discursive and structural leverage required by neoliberal governments and institutions wishing to extricate themselves from responsibility to annoying and belligerent constituencies, even if mass disaffection and indifference generally prevails. Organized networks are required to invent models of sustainability that go beyond the latest Plan of Action update, which is only then inserted into paper shredders of member states and "citizen friendly" businesses.

The empty center of neoliberalism is sociality. The organized network is part of a larger scramble to fill that void in trans-scalar ways.

On a more mundane, national, and local level, one only has to cast an eye toward the new legitimacy granted to the church as a provider of social "services." Civil society, in short, is replacing the ground of the social. But the assertion of the social is underpinned by ongoing antagonisms. The rise of right-wing populism is an example of how open the empty center is to a tolerance of fundamentalism.

Libertarian Legacies

Organized networks have their own problems to confront. Because of the lack of transparency about who is in charge of operations and project development, they are considerably slowed down. This is also a question of software architecture – the fact that we can't vote every month for who is the moderator for the month. There's no technical reason why we don't have this. Rather, it points again to the culture of networks – these can change fast in terms of applications, but not in terms of ideologies. To illustrate these issues, we'll turn now to a discussion of blogs, wikis, and Creative Commons.

The blog is a technology of networks, emerging in the first decade of the 21st century. Here the logic is that of the link. The link enhances visibility through a ranking system. This is how the blogger tackles the question of scale. But the question of scale cannot be reduced to one of scarcity. The linking technics of the blog don't add up to what we're calling organized networks. The blogger does not have infinite possibility but is governed by a moment of decision. This does not arise out of scarcity, since there is the ability of machines to read other machines. Rather, there are limits that arise out of the attention economy and out of affinity: I share your culture, I include your blog in my blogroll, or not, I don't share your culture, or I do; I like you, I don't like you. Here we see a new cartography of power that is peculiar to a symbolic economy of networks: recommendation for the few.

Quite importantly, the decisionism of the link constitutes a new field of the political. This is where schizo-production comes to an end. The naïve 90s Deleuzomania would say everything connects with everything. Technically speaking there's no reason why you can't include all the links of the world – this is what the Internet Archive does. The blog, however, is unable to do this – not due to a lack of space, since space is endlessly extensive through the logic of the link. Nor is this really an issue of resources. Instead, it is an issue that attends the enclave culture of blogs. They are zones of affinity with their own protectionist policies. If you're high up on the blog-scale of desirable association, the political is articulated by the endless requests for linkage. These cannot all be met, however, and resentment if not enemies are born. The enemy is always kept on the outside. They remain invisible. As such, the blog is closed to change. Blogs can thus be understood as incestuous networks of auto-reproduction. But they were highly labor intensive at the level of maintenance. It is therefore no real surprise that blogs faded out and have largely disappeared with the rise of social media networks such as Twitter and Instagram that were able to aggregate multiple feeds into the logic of the platform.

Since organized networks comprise new institutional forms whose relations are immanent to the media of communication, we can say that ultimately the blog does not correspond with the organized network. The outside for organized networks always plays a constitutive role in determining the direction, shape, and actions of the network. This is not the case for the blog, where the enemy is never present, never visible, since the network of the blog is the link, and the link is the friend.

Having said this, why did the blog command a degree of visibility in the mainstream media in a way that the organized network has not? Blogging started as a commentary on the mainstream media: TV, newspapers, and their websites. At a discursive level the blog was operating internally to mainstream media. In a genealogical sense the blog was part of the news industry. The main controversy within the news industry was whether or not bloggers could be considered as qualified journalists. This is part of a broader problem of categorization of the blogger: they are not poets, writers, scholars, etc. For a few years the blogger had become a profession with a professional code of ethics and job description, yet they still worked in conditions we associate with post-Fordist flexible labor. Paradoxically, then, the blogger was expunged and questioned by the networked organization even as it became the model subject of news generation.

The deep necessity or precondition of the blogger is not so much their networking capacity, since they are performing the self. Networking is secondary. But if you had a blogger who is self-performing without linking, they would remain invisible. Without the link you are non-existent. Thus their self-performance is identical to linking. However, there is a difference between networking and linking. There is a strong social network amongst bloggers, one that is highly intimate and highly disclosing of personal details. In that sense we can see a correspondence between the blog and reality television – the latter of course, is pretty much completely opposite to the logic of networks. So in terms of remediation, to what extent does this anti-networking character of reality TV carry over to blogs?

This is where we need readdress the idea of the political. As we have noted, with the blog, the political corresponds with the moment of linking, which is technically facilitated by the software, how it works, and the decisions that need to be made. Just as the blog is a self-performance, so too is the instantiation of the political. Both are an invisible undertaking. The fact that I do not link to you remains invisible. The unanswered email is the most significant one. So while the blog has some characteristics of the network, it is not open, it cannot change, because it closes itself to the potential for change and intervention. With the blog, you can comment but you cannot post. Your comments might even be taken down.

The blog, along with other now largely vanished social networks such as Friendster, Orkut, and MySpace, is finally characterized in terms of the software that refuses antagonisms. The early version of Orkut had a software interface that cut straight to the issue: "are you my friend? Yes/No." Only very few have the courage to tell someone straight in the face: "No." Seriously, what choice is there here, except to create an inflation of friends? We all want them. We find ourselves back to the 17 stages of joy. Nirvana Land. This is New Age revivalism at work, desperately insecure, and in search of a "friend."

The wiki offers another example of organized networks with its own specific social-technical characteristics. Here a collective intelligence is created, produced as a resource immanent to the media form. Yet it's important to understand that the wiki model will not work in all cultures and countries. The wiki is specific. It is a collaborative operation. You can have as many ideas as you want but this doesn't mean they will translate into a resource. The technical facilities on their own will not explain the story. Japanese and Chinese cultures, for example, do not like full visibility: to be seen, heard, or read. Why would they collaborate on these projects? Then think of the political histories of countries. The wiki presumes there is a willingness to work in the public and share knowledge. These are not universal values or aspirations.

The key to networks is the tension between open and closed systems of communication, ideas, and action. For the most part, e-democracy folk are unreconstructed techno-libertarians. The Creative Commons movement are also caught up in this persona, as if it's still 1999. Advocates of the Creative Commons license all too frequently claim they are "not political," as if this gesture will somehow enamor them to old-style institutions and publishing industries they are seeking to coax over to the other side. There is a naïve assumption that if Creative Commons can dissociate itself from leftist movements in particular, then they will have greater success in promoting Creative Commons as a dominant alternative to the strictures of intellectual property regimes. There is, however, no escape from politics, and the libertarian ethos of Lawrence Lessig and his cohorts would do well to be clearer about this.

The rhetoric of openness, shared by advocates of Creative Commons and libertarians, has purchase on governments who also trade in political populism.[5] Yet it disguises the political motivations and economic interests at work in these projects. The libertarian geek elite has so far effectively stopped networks from mobilizing their own financial resources. Most famously, there is the inability of networks to effectively work with micro-payment systems in the form of membership fees, software, etc. The libertarian geek option gives you one option: you give everything away for nothing and we'll take the money. Academic databases are an exception, where content (business data, reports, articles, etc.) can be accessed for substantial subscription fees. Institutions are fine with this arrangement, and don't seem too concerned about subsidizing these information services and publishing industries. The telcos also do okay – it's the poor hackers, activists, artists, and amateur intellectuals that get burnt.

The provocation of organized networks is to unveil these mechanisms of control and contradiction, to discuss the power of money flows, and to redirect funds. The organized network struggles with its own informality. This isn't a case of wanting a piece of the pie – organized networks don't even get a taste. No, organized networks want the whole bloody bakery! They are not examples for the network economy. Even in the case of Creative Commons, which do have a beta model of redistributing finance, this in fact is incredibly retrograde since it multiplies the necessity of intermediaries – a function eradicated in post-Fordist economies. You cannot earn money from content, only provide services

---

5   For a strong critique of openness, see Nathaniel Tkacz, *Wikipedia and the Politics of Openness* (Chicago: University of Chicago Press, 2015).

around it. In this nineties model of an information economy, the thing itself borders on being an untouchable sacred object, despite its banality. Again: the organized network has to break with the "information must be free" logic in order to move towards sustainability.

New Institutional Forms

Organized networks compete with established institutions in terms of branding and identity building, but it is as sites of knowledge production and concept development that primarily defines the competitive edge of organized networks. These days, most bricks and mortar institutions can only subtract value from networks. They are not merely unwilling but in fact incapable of giving anything back. Virtual networks are not yet represented in negotiations over budgets, grants, investments, and job hiring. At best they are seen as sources of inspiration amongst peers. This is where the real potential of virtual networks lies – they are enhancement engines. When they work well, they can inspire new expressions, new socialities, new technics.

The organized network is a hybrid formation: part tactical media, part institutional formation.[6] There are benefits to be obtained from both these lineages. The clear distinction of the organized network is that its institutional logic is internal to the social-technical dimensions of the media of communication. This means there is no universal formula for how an organized network might invent its conditions of existence. There will be no "internationalism" for networks.

While we have outlined the background condition of neoliberalism as integral to the emergence of organized networks, it also has to be said that just as uneven modernities created vastly different social and national experiences and formations, from the East to the West, from North to South, so too does capital in its neoliberal phase

---

6   Jeff Juris describes similar tensions between what he terms "horizontals" (self-organizing activist movements) and "verticals" (traditional institutions) as they played out across the various Social Forums in recent years. In reality, all forms of techno-sociality combine both horizontal and vertical forms of organization. Our argument is not so much that a hard distinction separates these modes of organization so much as a degree in scale. See Jeffrey S. Juris, "Social Forums and their Margins: Networking Logics and the Cultural Politics of Autonomous Space," *ephemera: theory & politics in organization* 5.2 (2005): 253–72, HTTP://WWW.EPHEMERAWEB.ORG/JOURNAL/5-2/5-2JURIS.PDF.

manifest in a plurality of ways. The diversity of conditions attached to free-trade agreements is just one example of the multiple forms of capital. From the standpoint of analysis, the understanding of capital is always going to vary according to the range of inputs one defines as constituting the action of capital. Similarly, no two organized networks will develop in the same manner, since their conditions of emergence are always internal to the situation at hand.

Eventually organized networks will be mirrored against the *networked organization*. But we're not there yet. There will not be an easy synthesis. Roughly speaking, one can witness a "convergence" between the informality of virtual networks and the formality of institutions. This process, however, is anything but harmonious. Clashes between networks and organizations are occurring before our very own eyes. Disputes condition and are internal to the creation of new institutional forms. Debris spreads in every possible direction, depending on the locality. The networked multitude, one could say, is constituted – and crushed – as a part of this process. In this sense, a new political subject is required, one that emerges out of the current state of disorganization that defines the multitude or connected masses. It is naïve to believe that, under the current circumstances, networks will win this battle (if you want to put it in those terms). This is precisely why networks need their own form of organization. In this process they will have to deal with the following three aspects: accountability, sustainability, and scalability.

Let's start with the question of who networks represent, or if indeed they hold such a capacity, and what form of internal democracy they envision. Formal networks have members but most online initiatives don't. Let's face it. Networks disintegrate traditional forms of representation. This is what makes the question "Did blogs affect the 2004 US-election?" so irrelevant. The blogosphere, at best, influenced a handful of TV and newspaper editors. Instead of spreading the word, the Net has questioned authority – any authority – and therefore was not useful to push this or that candidate up the rating-scale of electoral appeal. Networks that thrive higher up will eventually fail because they will be incorporated and thus degenerate into the capitalist mainstream. No matter what you think of Derrida, networks do not deconstruct society. It is deep linkage that matters, not some symbolic *coup d'état*. If there is an aim, it would be to parallel hegemony, which can only be achieved if underlying premises are constantly put under scrutiny by the initiators of the next techno-social wave of innovations.

The rise of "community informatics" as a field of research and project building in the mid-90s could be seen as an exemplary platform that could deal with the issues treated here.[7] Yet for all the interest community informatics has in building projects "from below," a substantial amount of research within this field is directed toward "e-democracy" issues. It is time to abandon the illusion that the myths of representational democracy might somehow be transferred and realized within networked settings. That is not going to happen. After all, the people benefiting from such endeavors as the World Summit of the Information Society (2003–2005) were, for the most part, those on the speaking and funding circuits, not people who are supposedly represented in such a process. Networks call for a new logics of politics, one based not just on a handpicked collection of NGOs that have identified themselves as "global civil society."

Networks are not institutions of representative democracy, despite the frequency with which they are expected to model themselves on such failed institutions. Instead, there is a search for "non-representational democratic" models of decision making that avoid classical models of representation and related identity politics. The emerging theme of non-representative democracies places an emphasis on process over its after-effect, consensus. Certainly, there's something attractive in process-oriented forms of governance. But ultimately the process model is about as sustainable as an earthworks sculpture burrowed into a patch of dirt called the 1970s. Process is fine as far as it integrates a plurality of forces into the network. But the primary questions remain: Where does it go? How long does it last? Why do it in the first place? But also: who is speaking? And: why bother? A focus on the vital forces that constitute social-technical life is thus required. Herein lie the variability and wildcards of organized networks. The persistence of dispute and disagreement can be taken as a given. Rational consensus models of democracy have proven, in their failure, that such underlying conditions of social-political life cannot be eradicated.

---

7   One of the many crossovers between computer science and humanities, as proposed by the late Michael Gurstein and others. Some of their texts can be found at HTTP://WWW.NETZWISSENSCHAFT.DE/SEM/POOL.HTM.

4.

URGENT APHORISMS

Four Stages of Web 2.0 Culture: Use. Modify. Distribute.
Ignore.
– Johan Sjerpstra

In between the blog posting and the tweet there is the aphorism, a centuries old literary form that should do well amongst creative media workers. Zipped knowledge of the 21ˢᵗ century.

Already for 18th century German experimental physicist and man
of letters, Georg Christoph Lichtenberg, there was an impossibility
for knowledge to capture the totality of things. "It is a question in
arts and sciences whether a *best* is possible beyond which our understanding cannot go" (Lichtenberg). The answer to Twittermania is not
the thousand page magnum opus. Today, in a techno-culture where
the link never ends, there is a need to give pause to thought. This is
the work of the aphorism. Karl Kraus: "An aphorism doesn't have to
be true. The aphorism should outstrip the truth, surpassing it in one
sentence." To all creative workers, migrants, vagabonds, activists, intellectuals of this world: Abandon the state, create multiple expressive

forms, engage in trans-border relations (affective, intellectual, social, political), invent new institutional forms!

Where to situate the study of network cultures? It hovers between a public form of "mass informality" and hard-core techno-determinism. The social noise we see scrolling down our screens is a waste product of techno-settings in which our sweet entries are situated. Interface is King, with the consequence that real techno-aesthetic intervention increasingly becomes a lost archive in the history of network cultures.

In retrospect Friedrich Kittler's techno-determinism remained an unfinished project. Kittler's post-1968 German media theory has not gone through many alterations since the early 1990s. The once bold statement "media determine our situation" doesn't shock anyone these days and has become an empty phrase. The media *a priori* is so obvious that it seems to have drifted into the realm of the collective unconscious. Henceforth no Kittler school. The grownup Kittler-*Jugend* are dedicated to scattered projects on the margins of academia. People once again obsess over their small careers and seem to have forgotten the primal energy that collective imagination can unleash. New generations read German media theory with interest but simply no longer have the time to read the necessary libraries to fully enjoy the details. Kittler himself abandoned contemporary techno-analysis and retired in imaginary Old Greece. How can there be a critique when such a position itself is still obscure and on the brink of disappearing? You start to sympathize with the programmer geeks when techno-determinism is sublimated by the highly attractive commercial sheen of platform capitalism.

Why network? We ought to ask this question. Why is the network, this empty signifier, the emerging-becoming-dominant paradigm of our age? Most of us will grow into network(ing) like children grow in and out of clothes. It takes some time to realize that we dedicate fixed periods of the day to the social-technical networks that are out there without factoring it in. Networking and communicating through email, chats, Twitter, and social networking sites are technological forms of day dreaming, a sphere you enter into and then come out of. The dreamtime in the techno-cloud could be compared to the siesta at the village square or chats in the local bar. It is time dedicated to the social. But what we get out of it is diffuse and impossible to quantify.

Why organize(d) networks? Organized networks are just one of many possibilities. But if the tendency that networks, over time, will simply have to become more structured, then why bother? Long live techno-social determinism. The org.net question should be preceded with: Why do we still talk about organization in an era that seems to celebrate looseness and non-commitment? *The Organization Man*, written in 1956 by William H. Whyte, is alive and well to this day. He did not disappear with the so-called end of industrialism. In fact, his powers have multiplied even if his "mind and soul" is no longer exclusively beholden to the demands of The Organization. Today, Organization Man has moved beyond that institutional terrain and penetrated the life of networks. Everyone is Organizing. Such was the great masterplan of the "organizational complex" (Reinhold Martin). Cooked up as a Cold War dream to extend the military-industrial complex into the realms of aesthetics and technology, the organizational complex fused the modulation of patterns from the Bauhaus School with the cybernetic programming of control.[1] "Media organize." This McLuhan-inspired maxim by Reinhold Martin truncates even further Friedrich Kittler's earlier synthesis, "media determine our situation." The key difference being the organizing capacity of communications media, which carry with it the organization man updated. This leaves us with the question: are we The Org Men? Wouldn't it be great to deconstruct the very .org concept to pieces in order to get rid of it, once and for all? Isn't there behind any call to organize a desire to restore the *über* organism once called tribe, church, society, nation-state?

The Tyranny of Failure

Not all online group initiatives work. Many fail. So can orgnets. The failure of a network is, however, not entirely without some work. There is always labor involved with failure. So we are using the notion of work in a different sense. We wish to invoke the idea of sustainability as a core feature of the work of networks. Failure is all too often the common of fragile conditions and the fragments of demands placed upon those involved in building and guiding the network. Social dust is a necessary precondition of the will to scale.

"We are here to stay." The sustainability issue is a highly political one. Once a network becomes sustainable it addresses the problem of

---

1   Reinhold Martin, *The Organizational Complex: Architecture, Media, and Corporate Space* (Cambridge, Mass: MIT Press, 2003).

time, which tends not to be allocated outside of the network sphere. More often networks are about the dimension of space – quite frequently, they are transnational in orientation. The material property of spatially distributed social-technical relations that are forever being remade through the logic of connection and speed provides sufficient grounds for distraction from the problem of time understood as the experiential condition of duration. This was the analysis of Canadian communications theorist and political economist Harold Innis, whose writings in the late 1940s and early fifties sought to address how it was that ancient civilizations rise and fall due to the spatial or temporal bias of their communications media and transport systems. The biases of our time are known to all, but ignored by even more.

"There ain't no time, only over time." The political aspect of networks is closely associated with the sustainability of time. The annoying network is the one that lasts the test of time and refuses to disappear. Networks as technoversity are connected to develop a diverse range of standards, practices, modes of governance, and social-technical relations. They collectively produce their own idioms of knowledge, one platform or system distinct from the next, all predicated on the will to communicate. The technoversity of networks is not simply about distribution across space but about maintaining lines of differentiation over time.

The realization of the social is no longer possible outside an understanding of the constitutive power of technologies. There is no pure social realm. The social is inseparable from the technology. We speak of healthy bodies and populations, but what is the healthy techno-social body? Why are fluidity and transformation such celebrated values these days? How can we design the care of the self for a social-technical network?

With so much real concern around ecological futures, how come there is so little concern within networks of techno-social futures? The net-cultural preoccupation with immediacy works against both the histories of the present as well as present conditions of the future. Network cultures have their own distinct apparatus of capture: Respond, Now! To cleave time from the work of networks requires a certain act of refusal through the practice of delay or, if you happen to be a member of the techno-economic elite, you simply log off. But these are not options for the networked masses. How, then, to reinvent a politics of autonomy in the time of networks? Such work requires new modalities of organization whose ambition is singular: conspire to invent new institutional forms.

Networks are not renowned for their managerial efficiency. Indeed, the very term "management" is one that makes many within networks actively hostile and they recoil with deep distaste. Networks are more inclined toward anti-authoritarian tendencies. They "unmanage" their cultural formation with little interest in purpose-driven, performance indicators and procedural guidelines. And it's no wonder they do this. Such practices are embedded in the highly dysfunctional audit cultures of dominant institutions. Networks are not goal driven. They galvanize around shared issues and the production of passions and the cultivation of memes, threads, rumors, and the like. The network blurs all purpose. That's why we raise the question of management in terms of organization. There can be no successful managerial science for networks. Please listen, once and for all, you brothers and sisters in consultancy land. Shy away from top-down decisions and impulses driven by regulatory *ressentiment*. IT-administrators belong in that category – their burning ambition is to ensure that networks never work.

Crisis as a Condition

Organized networks are best understood as new institutional forms whose social-technical dynamics are immanent to the culture of networks. Orgnets are partly conditioned by the crisis and, in many instances, failure of primary institutions of modernity (unions, firms, universities, the state) to address contemporary social, political, and economic problems in a post-broadcast era of digital culture and society. In this sense, organized networks belong to the era and prevailing conditions associated with post-modernity. Organized networks emphasize horizontal, mobile, distributed, and decentralized modes of relation. A culture of openness, sharing, and project-based forms of activity are key characteristics of organized networks. In this respect, organized networks are informed by the rise of open source software movements. Relationships among the majority of participants in organized networks are frequently experienced as fragmented and ephemeral. Often without formal rules, membership fees, or stable sources of income, many participants have loose ties with a range of networks.

The above characteristics inevitably lead to the challenge of governance and sustainability for networks. It's at this point that networks start to become organized. With a focus on the strategic dimension of governance, organized networks signal a point of departure from the short-termism and temporary political interventions of tactical media. At first glance orgnets are a natural, almost inevitable development of the "network society" as described by Manuel Castells. Yet nothing is "natural" in virtual environments. Everything needs to be constructed. And if so, under whose guidance? Who sets the very terms under which networks will cultivate their roots into society? Will this process of institutionalization have a (built-in) financial component?

As a political concept, organized networks provide what urban theorist Saskia Sassen calls an "analytical tool" with which to describe "the political" as it manifests within network societies and data economies. The social-technical antagonisms that underscore the political of organized networks are instantiated in the conflicts network cultures have with vertical systems of control: intellectual property regimes, system administrators, alpha-males, a tendency toward non-transparency, and a general lack of accountability.

How to rebuild labor organizations in the network society? This was one of the many unrealized ambitions of the anti- and later alter-globalization movements. And, for the most part, the unions never quite realized that life and labor within a digital paradigm had become the norm. Let us sketch out some of the current conditions challenging political organization within network societies. First, we need to problematize the concept of labor when understood as some kind of coherent, distinct entity. We know well that labor in fact is internally contradictory and holds multiple, differential registers that refuse easy connection (gender, class, ethnicity, age, mode of work, etc.). This is the problem of organization. How to "organize the unorganizables," to borrow from the title of Florian Schneider's documentary film. Second, we need to question the border between labor and life – contemporary bio-politics has rendered this border indistinct. Techniques of governance now interpenetrate all aspects of life as it is put to work and made productive. The result? No longer can we separate public from private, and this has massive implications for how we consider political organization today. What, in other words, is the space of political organization? Paolo Virno, for instance, speaks of a "non-state public sphere." But where, precisely is this sphere? All too often it seems networked, and nowhere. This is the trap of "virtuality," understood in its general sense. Of course there can be fantastic instances of political organization that remain exclusively at the level of the virtual, which is the territory of today's info-wars. Here, we find the continued fight over the society of the spectacle. Yet the problem of materiality nonetheless persists, and indeed becomes more urgent, as the ecological crisis makes all too clear (although this too is a contest of political agendas played out within the symbolic sphere).

Slogans: "R" Us – T-shirt label: Made for Asia – Today Your Friend, Tomorrow the World – Book title: "Stimulus and Indifference" – Praise Exodus, Blast Decay – Support My Exit – "Children of the Deconstruction" – The Institution is the Message – Project: Deleting Europe – I Joined the barcamp on anticyclic resistance and all I got is this lousy USB stick – Ethics is moral punk– Romantic Mobility – Silicon Friends™ – The Art of Attack (3 days intensive) – Post-Exotic: The Boring Other as Kulturideal – Buy More Consume Less – "Networking is Great to Waste Time Before Dying" – Rejected EU proposal: "Dialectics of Innovation: Creative Warfare in the Age of the Relaxed Crisis"

The Last Intellectuals

There are benefits in adopting a combinatory analytical and methodological approach that brings the virtual dimension of organization together with a material situation. This may take the form of an event or meeting, workshops, publishing activities, field research, urban experiments, migrant support centers, media laboratories ... there are many possibilities. In Italy, *uninomade* and the media-activist network and social center ESC are good examples of what we are talking about here. The Sarai media lab in New Delhi would be another. In the instance of bringing many capacities together around a common problem or field of interest we begin to see the development of a new institutional form. These institutions are networked, certainly, and far from the static culture and normative regimes of the bricks and mortar institutions of the modern era. Their mobile, ephemeral nature is both a strength and a weakness. The invention of new institutional forms that emerge within the process of organizing networks is absolutely central to the rebuilding of labor organizations within contemporary settings. Such developments should not be seen as a burden or something that closes down the spontaneity, freedom, and culture of sharing and participation that we enjoy so much within social networks. As translation devices, these new institutions facilitate trans-institutional connections. In this connection we find multiple antagonisms, indeed such connections make visible new territories of the political.

Reading Russell Jacoby's *The Last Intellectuals: American Culture in the Age of Academe* (1987) two decades later makes you wonder how such an independent study would look like, post-Cold War, post-9/11, in the age of the internet and globalization.[2] Jacoby's description of the "impoverishment of public culture" has not come to a halt. No dialectical turn here. As predicted, the figure of the "public intellectual" has disappeared. "Intellectuals no longer need or want a larger audience; they are almost exclusively professors who situate themselves within fields and disciplines." The non-academic intellectuals, an endangered species in the 1980s, have vanished for good. The academics who replaced the general intellectuals created "insular societies." There is a widespread fear here of the "single-minded men." But are we really living in the Age of the Expert? It is not the expert knowledge that has become the dominant voice in the media age. Instead, we have witnessed the rise of the celebrity, and the "celebrification" of all spheres of mediated life. The professional is hiding inside the walls of the office culture. Instead of a Triumph of the Professional we witness the Cult of the Amateur (Andrew Keen), neither of them claiming any of the virtues of the General Intellectual. Nothing in Jacoby's study points at the appearance of "citizen journalism," "participatory culture" (Henry Jenkins), and the decline of professional work due to the rise of free content found in free newspapers and through the internet. Yesterday's public intellectuals of mass media were not exactly unpaid fellow travelers. What would Jacoby's strategy be after the "de-monetarization" of the media markets?

Communication conditions the possibility of new political organizations. We could say that the political of network societies is comprised of the tension between horizontal modes of communication and vertical regimes of control. Just think of the ongoing battles between internet and intellectual property regulators such as WIPO (World Intellectual Property Organization) and pirate networks of software, music, or film distribution. Collaborative constitution emerges precisely in the instance of confrontation. In this sense, the horizontal

---

2   Russell Jacoby, *The Last Intellectuals: American Culture in the Age of Academe* (New York: Basic Books, 1987).

and vertical axes of communication are not separate or opposed but mutually constitutive. How to manage or deal with these two axes of communication is often a source of tension within networks. Here, we are talking about models of governance, without universal ideals to draw on. More often than not, networks adopt a trial-and-error approach to governance. It is better to recognize that governance is not a dirty word, but one that is internal to the logic and protocols of self-organization.

Participation Economies and Free Labor

The "participation economy" of Web 2.0 and platform capitalism is underscored by a great tension between the "free labor" (Tiziana Terranova) of cooperation that defines social networks and its appropriation by firms and companies. How is the "wealth of networks" (Yochai Benkler) to be protected from exploitation? Unions, in their industrial form, functioned to protect workers against exploitation and represent their right to fair and decent working conditions. But what happens when leisure activity becomes a form of profit generation for companies? Popular social networking sites such as Facebook, MySpace, Bebo, del.icio.us, and the data trails we leave with Google function as informational gold mines for the owners of these sites. Advertising space and, more importantly, the sale of aggregated data are the staples of the participation economy. No longer can the union appeal to the subjugated, oppressed experience of workers when users voluntarily submit information and make no demands for a share of profits. Nonetheless, we are starting to see some changes on this front as users become increasingly aware of their productive capacities and can quickly abandon a social networking site in the same manner in which they initially swarmed toward it. Companies, then, are vulnerable to the roaming tastes of the networked masses whose cooperative labor determines their wealth. This cooperative labor constitutes a form of power that has the potential to be mobilized in political ways, yet so rarely is. Perhaps that will change before too long. Certainly, the production of this type of political subjectivity is preferable to the pretty revolting culture of "shareholder democracy" that has come to define political expression for the neoliberal citizen.

The precarity debate was, correctly, about the material conditions of labor and life. Mistakenly, the precarity discourse remained fixated on the rear-view mirror of Fordist production and the welfare state.

But there is more to precarity than this. Judith Butler wished to extend the term to include emotional states and affective relations. Yet somehow precarity doesn't satisfactorily capture the intensity – and dullness – of the contemporary soul. What comes closer is the image of the nervous, electric body in the late nineteenth and early twentieth century as diagnosed in sociological accounts of urban transformation. Think Georg Simmel, Gabriel Tarde, Walter Benjamin. The image of digital disembodiment was perhaps a 1990s attempt to update the electric body, but nowadays such a notion just looks sadly comical and misplaced, which brings us back to the materiality of communication vis-à-vis Kittler. Today we have not so much digital disembodiment but the violence of code that penetrates the brain and the body. It is the normality of difference, sending out constant semiotic vibrations, that numbs us. What the precarity meme doesn't catch is the cool frenzy. There is an aesthetics of uncertainty at work. An impulse to Just Do It! Extreme Sports. Risk Societies. Financial Derivatives. Creative Classes. Porn Stars. Game Cultures. Today, it seems impossible to escape the network paradigm that is always economically productive, even if it never returns the user a buck. The non-remunerated body remains a body in labor. And it is increasingly exhausted. The brain encounters the limits of the day and everything that is left uncompleted. The endless task of chores ticked off slide over from one day to the next. One becomes tired by looking at the "to-do" list, which reproduces like a nasty virus. Bring on the remix.

The shift from Fordist modes of assembly production to post-Fordist modes of flexibilization cannot be accounted for by reference alone to capital's demands for enhanced efficiency through restructuring and rescaling. The 1970s in Italy saw the rise of *operaismo* (autonomist workerism) who refused the erosion of life by the demands of wage labor. Importantly, their unique "refusal of labor" demonstrates, in theory, a clear capacity of workers to change the practices of capital, for better and worse. The Italian collective strike is a one-off concept workshop, blending the radical with the general. It is in this power of transformation that "the common" is created (unlike so many other struggles and forms of dissent in Europe). The ongoing challenge remains how to organize that potentiality in ways that produce subjectivities that can open a better life – in Italy, and beyond.

Workfare, flexicurity, a universal basic income, or "commonfare" – all of these options are variations on the theme of state intervention that is able to supply a relative security to the otherwise uncertainty of labor and life. Such calls are misguided. They presuppose that somehow the state resides outside of market fluctuations and the precarity of capital. The state is coextensive with capital. The 2008 credit crisis has shown the state has little command over the uncertainties of finance capital. How, then, can the state guarantee stability? Furthermore, to whom does the state offer security? Certainly not to undocumented migrants. The call for flexicurity is a regressive, nostalgic move that holds dangerous implications vis-à-vis the formation of zones of exclusion. There is no pleasure principle in being underpaid. The price of freedom is a high one and it is only a handful of lucky outsiders in the Rest of the West who can afford to work for free, enjoying unemployment while living off a small income. It is a secret lifestyle choice for a diminishing elite of cultural conceptualists and their outsourced army of semiotic producers. This is not what the dreams of the multitudes aspire to realize. There is much political value in targeting not the state but the companies – especially those engaged in the extractivist data economies – and insisting on a distribution of income commensurate with the collective labor that defines the participation economy.[3] This may be a more effective strategy for broadening the constitutive range of labor organizations.

If social movements are serious about addressing the economic conditions of workers and engaging the complexities of the political they would put an end to the mistaken faith in the state as the source of guarantees. Moreover, the logic of the state as a provider of welfare is special to Europe – it does not translate to the situations of workers in many Asian countries, for example. So what are the borders of connection among workers? Does the movement simply reproduce the borders of the EU? Or does it engage in the much harder but no less necessary work of transnational connection? If so, then targeting the state does not especially help facilitate a common territory of organization. The global circuits of capital are where radical politics should focus their attention. But global capital is in no way uniform in its effects, techniques of management, or accumulative regimes. Political intervention, in other words, must always be situated while

---

3   On extractivism, see Sandro Mezzadra, Sandro and Brett Neilson, "On the Multiple Frontiers of Extraction: Excavating Contemporary Capitalism," *Cultural Studies* 31.2-3 (2017): 185–204.

traversing a range of scales: social-subjective, institutional, geocultural. The movement of relations (social, political, economic) across and within this complex field of forces comprises the practical work of translation. Translation is the art of differential connection and constitutes the common from which new institutional forms may arise.

Practices of collaborative constitution are defined by struggle. There is no escape from struggle and the tensions that accompany collaborative relations. This is the territory of the political – a space of antagonism that in our view is much more complicated than the friend/enemy distinction of Carl Schmitt. Again, it is the work of translation that reveals the multiplicity of tensions. As Naoki Sakai and Jon Solomon have written, translation is not about linguistic equivalence or co-figuration, but rather about the production of singularities through relational encounters. But let's get more concrete here. What is a relational encounter? It occurs through the instance of working or being with others. Of sharing, producing, creating, listening. Sustaining a range of idioms of experience is a struggle in itself – one that is rarely continuous, but rather continually remade and reassembled. This in turn is the recombinatory space and time of new institutions.

Let's unpack the idea of new institutions and their relation to precarity. If we say that precarity and flexibility is the common condition – one that traverses class and geocultural scales – then we can ask: what is the situation within which precarity expresses itself? The situation (concept + problem) will define the emergence of a new institution. Situation, here, consists of virtual/networked, material, affective, linguistic, and social registers. We are of course always in a situation, but how to connect with others? The point of connection brings about tensions – the space of the political – and the ensemble of relations furnishes expression with its contours. Real power lies not in the spectacle of the event, but rather subsists within the resonance of experience and the minor connections and practices that occur before and after the event. That is the time and space of institution formation. The rest is a public declaration of existence.

The Organization Man

The question of organization persists: Who does it? How is organization organized? For Keller Easterling, this is the role of the orgmen: "Different from the deliberately authored building envelope, spatial products substitute spin, logistics, and management styles for considerations of location, geometry, or enclosure. The architect and salesman of such things as golf resorts or container ports is a new orgman. He designs the software for new games of spatial production to be played the same way whether in Texas or Taiwan. The coordinates of this software are measured not in latitude and longitude but in the orgman argot of acronyms and stats – in annual days of sunshine, ocean temperatures, flight distances, runway noise restrictions, the time needed for a round of golf, time needed for a shopping spree, TEUs, layovers, number of passengers, bandwidth, time zones, and labor costs. Data streams are the levers of spatial manipulation, and the orgman has a frontier enthusiasm for this abstract territory. He derives a pioneering sense of creation from matching a labor cost, a time zone, and a desire to generate distinct forms of urban space, even distinct species of global city."[4]

The OrgMen of networks, then, share something with the alpha-males and sysops (system operators). Both administer behaviors in symbolic or technical ways, shaping patterns of relation. Indeed, the software architecture used by any network is its own org(wo)man. Organized networks would do well to diversify their platforms of communication, adopting a range of software options to enable the multiplication of expression and distribute as much as possible the delegation of network governance. If one platform starts to fall flat – say a mailing list – then perhaps the collective blog is going to appeal to more. Whenever the collective labor of a network can be galvanized around forms of coproduction (making an online journal, organizing an event, setting up a file-distribution system, producing a documentary, identifying future directions, staging a hack, designing slogans) then the life of the network finds that it has a life. Such techniques of collaborative constitution keep in check the proto-fascistic tendencies of the orgman that lurks within every network. The tension between these two registers of network sociality is a necessary dynamic. The challenge is to keep the game in play, gradually shifting the limits of the network disposition.

If we were to reinvent cybernetics (as an organizing logic of recombination, feedback, noise, etc.), outside the military-industrial context

---

4   Keller Easterling, *Enduring Innocence: Global Architecture and its Political Masquerades* (Cambridge, Mass.: MIT Press, 2005), 2.

of the Cold War, what would it be? First of all, it would no longer be obsessed with biology and bio metaphors. The aim of computer networks is not to mimic the human by copying or improving human features such as the brain, memory, senses, and extensions. The question of agency and the relation between humans and non-humans, as thematized by for example Bruno Latour and the actor-network theory crowd, is a typical remainder of the cybernetics 1.0 era. In the past cybernetics tried to figure out how to connect the individual (human) body to the machine. It presumes we still have an issue with "intelligent machines." The cybernetic 1.0 age was both worried and drawn to the idea that the human can(not) be replaced by thinking machines. The result of this was an irrelevant debate for decades over artificial intelligence (AI). These days no one is concerned if and when the machines take over. Have you ever been scared by the idea that a computer can and will beat you at chess? Sure it can, but so what? We know Big Brother is storing all the information in the world. AI is here to stay but is no longer a key project in technology research. Whereas cybernetics 1.0 tried to schematize human behavior in order to simulate it through models, cybernetics 2.0 is concerned with the truly messy, all too human, social complexity. We are not ants. We are more and behave as less. Our understanding has to go beyond the boring mirror dynamic of man and machine. Computer science will have to make the leap into inter-human relations in the same way as humans are adapting to the limits set by computer interfaces and architectures. Stop the mimicry procedures, and restart computer science itself.

Reinhold Martin: "Wiener argues that what counts is not the size of the basic components (such as neurons, which are similar in humans and ants) but their organization, which determines the 'absolute size' of the organization's nervous system – its upper limit of growth and index of social advancement. An organism's social potential, conceived in terms of its ability to organize into complex communication networks, is thus measured as a function of the size of its internal circulatory and communications system, which is a function, in turn, of their own organizational complexity. The original analogy between the social and biological organism is thus collapsed, as the two become directly linked as part of the same network.... A relational logic of flexible connection replaces a mechanical logic of rigid compartmentalization, and the decisive organizational factor is no longer the vertical subordination of parts to the whole but rather the degree to which the connections permit, regulate, and respond to the informational flows in all directions."[5]

What are the limits of potentiality for the organized network? While impossible to answer in terms of content (every network has its own special attributes), we can say something here about form. Form furnishes the contours of expression as it subsists within the social-technical dynamics of digital media. How these relations coalesce as distinct networks situated within and against broader geopolitical forces becomes a primary challenge for networks desiring scalar transformation – a movement that also consists of trans-institutional, disciplinary, subjective, and corporeal relations whose antagonisms define the multiple registers of the political. The question of limits takes us to the trans-scalar practice of transversality – the production of multiple connections that move across a range of social, geocultural, and institutional settings. There are also strategic questions: Who do you collaborate with? How local are you? Are you willing to deal with the cynical professionals of traditional media? Do you believe in Meme Power, viral marketing, and subliminal dissemination with the chance of hitting the Zeitgeist lottery, or in the hard work of political campaigning?

Collaboration is always accompanied by conflict and struggle. This is a matter of degree. And there'll be plenty of exhilaration that keeps the momentum going. But tensions will always be present. Better to work out an approach to deal with this, otherwise you'll find your projects go kaput!

---

5   Martin, *The Organizational Complex*, 23.

5.

CONCEPT PRODUCTION

What does it reverse or flip into when pushed to the
limits of its potential?
– Marshall McLuhan, *Laws of Media*, 1988.

I have seen the future – and it's not visual.
– Johan Sjerpstra

DURING THE FIRST DECADE OF THE 21ST CENTURY THE ACADEMIC DIS-
cipline of media studies failed to develop a compelling agenda. Media
turned out to be empty containers, individualizing people rather
than imagining collective agendas. The growth of "media" could lead
to its ultimate implosion. If media have gone digital and become
the network glue between devices, there is a danger of defining the
boundaries of media studies purely for the sake of the discipline itself.
Media studies then becomes self-referential, defined solely in terms
of its self-defense against predatory competitors. For instance, if media cannot be distinguished anymore from urban life, geography, and

location-based services, then what is the task of media studies? What, indeed, is the object of media in an age of mediated environments?[1] Public Relations is a trap here: to study media is not identical to its promotion. We need media researchers to reflect on how they use their object of study in the research methodology itself. In a media society of compulsive immersion, this is no easy task. Indeed, many would charge such a call as regressive, harking back to the Cartesian myth of critical detachment. But as we will argue, we consider the work of reflexive mediation – of concept production – necessary if such a thing as media critique is to exist at all.

For the past decade media studies has struggled to keep up with the pace of techno-cultural change. The methodologies and concepts of the broadcast era of "mass media" are of little use when analyzing networked digital cultures. The globalization of higher education and the increasing competition between disciplines over diminishing funds and international students has further exacerbated the unconscious crisis of media studies. With a push towards vocational training, stagnating cultural studies, and a distaste for theory in general, film and television studies can only make defensive gestures towards the ever-expanding digital realm.

The future of media studies rests on its capacity to avoid forced synergies towards "screen cultures" or "visual studies" and instead to invent new institutional forms that connect with the trans-media, collaborative, and self-organizational culture of teaching and research networks. Unless media studies makes such a move, it will join the vanishing objects it assumes as constitutive of media in society. In this chapter we want to go beyond an inventory on the state of the art and use the example of *organizing networks* as a concept in development that might revitalize education and research in this field. The work of organizing networks involves the invention of new institutional forms immanent to communications media. Such a collaborative process mediated through network culture conditions the possibility of disciplinary transmutation.

Database Dating and the Invention of Schools

A school is defined as a critical mass of equal standing scholars and teachers with distinct intellectual traits. The question on the table is

---

1   See Sean Cubitt, *Finite Media: Environmental Implications of Digital Technologies* (Durham and London: Duke University Press, 2017).

how to develop decentralized "global" (public) research schools and identify what their key features might be. The current "science" system expects collaborative research but does not facilitate the formation of distinctive schools of concept production. Humanities scholars have learned to play the waiting game of drip-fed funding, locked in the holding bay of a terminal with no interest in supporting experimental research on a distributed, trans-institutional scale. The internet has, surprisingly, not made much of a difference. One could blame the structure of research funding and related academic publishing rituals for this, but that's a weak proposition since it assumes a form of structural over-determination. A general culture of indifference towards the relation between media environments and conditions of concept and knowledge production prevails in a more pervasive way. Has the (neoliberal) individualization of society increased our fear for long-lasting collective commitments? If "schools" are so influential, what stops us from creating them?

Instead of conceiving a multitude of schools and approaches that theorize the turbulent transformations in the media sphere, the collective production of concepts has been a decidedly tame affair. There are no institutional examples of new media research agendas, collectives, or schools. Where is the Frankfurt School of our time? All too soon we seem to reduce collaborative efforts to fashions in theory. No matter one's critiques of the Frankfurt School, the fact is they produced a lasting legacy in the field of media research and cultural analysis. Despite its identity as a school, the work of individuals stands out: Benjamin, Adorno, Horkheimer, Marcuse. But a counter-reading would say it is exactly the distributed intellectual labor of the Frankfurt School that allowed these extra-institutional figures to become part of the canon of critical theory and social philosophy. It is the self-organization of a group of intellectuals to create a school across national and transnational spaces that inspires us here.[2] Much of what is now called French theory has been produced in similar circumstances, outside or on the fringes of academia, creating a delicate balance between individual work and intense exchanges within a social milieu of intellectuals, artists, and activists. A similar story can

---

2   For a discussion of the Frankfurt School as a precursor to organized networks, see Ned Rossiter, *Organized Networks: Media Theory, Creative Labour, New Institutions* (Rotterdam: NAi, 2006), 19–22.

be told of Italian post-*autonomia* political philosophy and the earlier schools of Freudian psychoanalysis. Media studies desperately needs its own versions of these distinct, collective intellectual efforts.

Is global media studies something for the future? That is not the direction we would advocate, at least in terms of how it is currently defined. Global media studies emphasizes cultural competence and an awareness of cultural differences derived, in some cases, through media ethnography. For the most part, global media studies is driven by conservative methods of content and discourse analysis. Frequently it stays within the UK audience studies approach to media research and is, at best, Commonwealth in its focus. Certainly, there is research on the political economy of global media, though here the cultural question is frequently left out. And it plays into very weak empirical attempts at a political economy of so-called "global" media industries, which in reality are more often regional at best with the rise of "national webs" bringing even the regional scale of media-culture into question. In short, global media studies is a concept free zone without transnational connection or inter-scalar complexity. There is no coordinated network of global media researchers working in any sustained manner on the undulations of contemporary media cultures. The old association form of grouping international researchers around particular disciplines is certainly not the model that is going to provide the distributed labor required today for research practices immanent to the media of communication.

Traditionally renowned as the dream factory, the USA in particular has demonstrated an inability to create collective academic projects and facilities for long-term collaborative work in digital media research. In fact, the world over, the current research model is one of a "principal investigator" who hires assistants, postdocs, and PhD students with the sole aim to demonstrate "leadership." What characterizes media studies, in sync with the global neoliberal context, is the figure of the lone researcher. If research funds are awarded, content might consolidate within the logic of reporting and listing of outputs. But if the money doesn't come through it can be very hard to materialize research, at least within the academy. While a plethora of tactical media interventions over the past 10–20 years suggest a healthy state of media invention, these are not instances of collective production at a disciplinary level of institutional support, infrastructure, and commitment.[3] Rather, the academy's preference is for silo construction that engenders, at best, interdisciplinary innovation around intellectual property regimes (IPRs) as distinct from transdisciplinary invention through free collaboration, and remains the model of the singular author-producer.

Perhaps teamwork's violation of the sacred author as individual explains why science's collaborative methods are by and large not allowed in the humanities – despite the fact that the "science" model is promoted so heavily as a cultural model. The sciences long ago recognized the collaborative function of knowledge production, and a key part of this stems from science's old collaborations with industry, requiring the field to traverse institutional expectations and logics of production. But if we look at a case like Australia where "industry linkage" grants in the humanities have been the norm for over a decade, this structural change has not had any widespread sustained impact in terms of materializing meaningful collaborative research practices. In the end these are formal relations on paper and not lived relations in any substantive sense in terms of disciplinary invention. Moreover, they are top-down arrangements driven by government policy directives on obtaining funds within a populist and economistic political paradigm. It is no surprise that Aussie-academics drifted toward industry linkage grants out of simple cynical opportunism along with a desperate survival instinct, a few exceptions aside. The so-called industry partners are often just neighboring governmental bodies such as cultural and media organizations and government agencies. The "linkage" is largely symbolic and institutionally, not research, driven. Meaningful and inventive research collaborations cannot materialize in such arrangements when they start with these kind of forced relations.

The database approach to finding research partners within the European funding system is not much different. Academics in search of research partners plug in their expertise details, disciplinary training, and project interests. Press submit and the database builds a field of partner options. Within the European Framework Programmes

---

3   As examples of non-academic research collectives, one could mention groups like Critical Art Ensemble, Institute for Applied Autonomy, Bureau d'études, Beaver Group, and the European Institute for Progressive Cultural Policies, among many more. Our question is why the academy seems incapable of generating such inspiring and productive formats and practices of research.

(now replaced by the Horizon 2020 "research and innovation" programs), researchers from different countries and disciplines are forced to work together in order to have a shot at funding. In such endeavors the database serves as a mirror of TV dating games where contestant profiles are the basis upon which a "perfect match" is conjured from the abyss of isolation. It doesn't help that researchers are quietly advised to lobby in Brussels for collaborative EU research money if they wish their projects to have any chance of obtaining funding. With assessors holed up in hotels in Brussels reviewing a thousand applications over a few days, it's no wonder that they pick their buddies out of the stack. Such a massive scale of applications in the European system leaves little choice other than a corrupt culture of cronyism. While efforts in public relations might improve the signal/noise ratio with decision makers, it is very hard to see how it assists the work of collaborative research, assuming your project even has the capacity and resources to enlist an Ad Man. Instead, this model of research becomes indistinct from contemporary political campaigns, where special interest groups lobby the centers of power and all the while the project moves further away from the objects, methods, and passions of research.

In sum, the academy's closed intellectual property regime logic, its centralization, and its habit of database dating is thoroughly disconnected from digital media environments characterized by a culture of open communication and collaboration. Among countless current examples of this are open courseware (from MIT to Edu-Factory), open publishing (online journals, collaborative blogs, etc.), and open software movements (openstreetmap.org, Linux, and all its variants). Crucial here is the connection between online communication and material situations. Whether it's bar camps, collective field research, or group planning meetings, the offline element is central to the work of organizing networks. This is what building your own infrastructure means, which we now know the alt-right movement has been doing for years. Meanwhile the left continues to bury its head in the sand over this issue. To be purely "virtual" is to risk drifting potentially vital concepts into the realm of vaporware. The material dimension functions to galvanize online activities as a substantive force.

Successful instances of disciplinary collaborative research can exist. We are reminded of the science and technology studies (STS) method in which anthropologists and sociologists approached science with questions. Why did it work? Driven by initial questions around health, environmental, financial, and medical issues, STS was a research driven approach to knowledge production, very different from the self-referential tendency in media studies to avoid rigor and instead produce texts based on the endless citation of peers with power. STS provides an example of how the material world forced the disciplinary contours and methods of STS to adapt, evolve, and consolidate. STS managed to develop the skill of translating issues that emerge in society into legitimate and necessary research programs free from client driven research. While media studies all too often remains beholden to the abstract and frequently politically motivated research focus areas set by funding agencies, it could embrace the culture of collaboration special to networks and design new models of autonomous research funding.

The Media Question and Organized Networks

So let's face it, after a short renaissance of Anglophone cultural studies in the early 1990s, the Media Question no longer sparks the collective imagination. Abundant fragmentation has distracted us. Why do we even bother with the fate of media studies? What's really so special about it? Certainly we have our own intellectual affinity with various figures – Innis, McLuhan, Flusser, Kittler, Virilio, and so forth. But this says more about attractions to style and modes of analysis than disciplinarity *per se*. We would also have to admit to a fascination with the magic of mediated relations and the marvels of technical invention. But what drives us is the Media Question, still unresolved at the level of architectural control, the production of standards, and management of protocols – all things we see as collectively produced and in fact unrelated to the routine and very dull fare that usually defines the territory of media studies: representations of marginality, textual and content analysis, ethnographies of the dreary, and identity politics.

The direction of media studies must therefore be non-representational. While for the time being media remains what it has been, a container concept, we know from the work of medium theorists that the container is a very elastic form of the materiality of expression. Media never cease to surprise or generate unusually high levels of excitement and forms of mass participation. In this respect, media is always social, and so too is the materiality of form. The focus should, for instance, be on integrated urban-media environments, smart technologies, ubiquitous media. As a social-technical phenomenon, then, communications media are always alive and therefore changing.

In contrast, what makes media studies so boring and predictable is its slowness and obsession with the visual. No matter how fast the transmission of signals, the distribution of academic knowledge is inherently slow. Think of book publishing and how you wait two years for the text to appear. The cost of printing books in China is cheap, but shipping containers still take two to three months to reach the other side of the world. If key publishers embrace print-on-demand some of these obstacles would be addressed. And it is truly unforgiveable the way publishers such as Routledge and Palgrave charge upwards of one hundred pounds sterling for monographs. Authors should also know better and take a closer look at their contracts before they consign their hard work to the remainder's shelf and pulping machines at the expense of being read in a timely way.

Media studies hasn't taken its objects seriously enough. All too easily it borrows from the shopping mall of post-modern theory, psychoanalysis, sociology, and literary studies to interpret its own object, neglecting the challenge to develop media theory itself. The insane subject provides the raw material and object for concept production in psychoanalysis, for example. So why should it be so difficult for media technologies to function as the object of concept production? To think that it can't is to seriously regress and suppose that media are simply empty vessels or black boxes waiting to be filled – the container theory all over again – rather than remarkable objects with their own capacity for inspiring concepts. Where, then, should media studies find its concepts of renewal? What are the geocultural contours and institutional settings best equipped to support collaborative concept production that will most likely happen outside the academy?

Let us be clear: concepts are essential for a vibrant and distinctive media studies discipline. They are not produced for their own sake, but because they hold the capacity to transform society in profound ways. Think of the Freudian inspired work of Edward Bernays and the rise of public opinion and advertisements of consumer fantasies in the US after World War II.[4] The experience of life and the

---

4   For a fascinating analysis of Bernays, cybernetics, and the formation of neo-liberal governance, see Brian Holmes, "Adam Curtis: Alarm Clock Films," in *Escape the Overcode: Activist Art in the Control Society* (Eindhoven and Zagreb:

production of subjectivity underwent enormous change once the concept of desire was integrated into the media imaginary of post-War society. Consider also how the Obama 2008 presidential campaign drew heavily on both traditional grassroots political strategies along with organizational concepts developed out of social-network and activist media such as swarming, viral marketing, and self-organization, resulting in a post-party production of the Obama brand. Trump has merely amplified this technique. For these reasons, we think there is great importance in developing distributed, collaborative research agendas that invest in analysis by way of practice to produce concepts in the digital era.

One approach to addressing the need for new modes of association and collaboration in media studies is to experiment with the model of organized networks to institute media research within an informational, digital paradigm. Such a move would further disaggregate the already crumbling system of the modern university. Critics might contend that organizing media research within the culture of networks would simply reinforce the neoliberal drive to outsource the provision of services. One could also argue that a move toward organized networks would be precisely at odds with the idea of institutional security necessary for sustainable practices of knowledge production. But in our minds there's an important distinction between service provision and knowledge production. The former is driven by the market impulse and is underscored by disinterested ("alienated") labor within highly precarious conditions of employment. While the latter is not without similar precarity, and indeed in the age of "digital diploma mills" shares much with service provision,[5] the work of knowledge production in the age of digital networks nonetheless holds both affective and material qualities of a special kind, quite unlike so-called knowledge production within universities, where social relations are as hierarchical and atomized as those found on the 19th century factory floor.

Social relations within and across networks tend to be fleeting and project driven, struggling for continuity. Similarly, at a material level the technical standards of much online knowledge production tend

---

Van Abbemuseum Public Research / WHW, 2009), 284–303.

5   See David Noble, *Digital Diploma Mills: The Automation of Higher Education* (New York: Monthly Review Press, 2002).

to result in a proliferation of platforms and sites, which has the effect of further disaggregating what might in fact at the social level be a series of projects shared by a number of participants distributed across transnational spaces and times. This leads us to another key concept for knowledge production in the age of networks: seriality. Where standards, sites, and participants appear fragmented, the concept of seriality helps establish connections across otherwise seemingly unrelated phenomena. Tools and methods need to be developed that assist researchers, participants, and user-viewers in their efforts to keep track of multiple projects distributed across space and time. This requires more than a real-time software fix. Take Google Wave, for instance. According to the Wikipedia definition "Google Wave is a web-based, computing platform, and communications protocol, designed to merge key features of media like e-mail, instant messaging, wikis, and social networking. Communications using the system can be synchronous and/or asynchronous, depending on the preference of individual users."[6] The problem here is not so much the baroque *combinatoria* of data streams but the standardized automatic results these dashboards generate to instantly solve the all too human noise of collaboration. Aggregation is not the answer for meaningful collaboration, since the productive and indeed often charming differences that define research projects online tend to be erased when incorporated into a real-time aggregation platform. Networks thrive on not-working and communication gaps.[7] Special qualities of different communications media need to be retained rather than aggregated to best sustain serial relations across diverse networked projects that refuse incorporation by any single power, platform, or authority.

Concepts Beyond the Classroom

There also remains considerable work to be done concerning the relationship between research and teaching in media studies and indeed across the disciplines overall. Moreover, there is great potential to generate concepts in the research-teaching "nexus," to adopt the managerial parlance. But this work of concept co-production should not

6   HTTP://EN.WIKIPEDIA.ORG/WIKI/GOOGLE_WAVE.

7   Geert Lovink, *The Principle of Notworking: Concepts in Critical Internet Culture*, Public Lecture, Hogeschool van Amsterdam, February 24, 2005. Available at: HTTP://WWW.HVA.NL/LECTORATEN/DOCUMENTEN/OL09-050224-LOVINK.PDF.

be technologically determinist or have too much faith in technology alone. Like the "ICT for development" discourse, which assumes the introduction of technology into developing countries results in economic transformation, there is a broad feeling that technology makes teaching better. Nevertheless, the vast majority of academics in the US continue to resist adopting technology in the classroom.[8] For us the issue is not so much one of mass conversion to high-tech education (the students come to class pre-packaged with a digital default anyway, even if their professors do not), but of translating research practices from the wikis, collaborative blogs, project based learning, and media production that many of us already engage in for pedagogical purposes. In other words, the use of digital media in the classroom already assumes a process of research: one has to learn new software, techniques of information retrieval and recombination, social protocols of communication, and so forth. Moving such practices into a paradigm of teaching-as-research requires translation, which in this case is a collective political task across institutional settings that at once formalizes and consolidates teaching and research through the development of new methods, models, and concepts.

Instead, what we find within the techno-institutional economy is research excluding teaching, thereby retaining the distinction between the two. Moreover, media studies is positioned as a disciplinary practice of education rather than research. Within the circumscribed borders of the informational university, media studies has no distinct object of study. The object of media studies is not media, it is audiences, textual analysis, and content analysis coupled with a vocational element. Media studies for most students is a means to an end: the job. Analyzing the process of learning itself is very rarely embraced and requires a good teacher. Students, broadly speaking, are terrified by the sea of uncertainty that defines the labor of thinking and therefore research. They want signposts installed at every point of the learning process, and academics willingly supply students with ultimately false rules of certainty in preparation for good end of semester teacher performance evaluations required to hold on to the job. This situation of pedagogical failure and disciplinary stagnation is hardly helped by

---

8   See Jeffrey R. Young, "Reaching the Last Technology Holdouts at the Front of the Classroom," *The Chronicle of Higher Education*, July 24, 2010, HTTP://CHRONICLE.COM/ARTICLE/REACHING-THE-LAST-TECHNOLOGY/123659/.

academics whose sole aspiration at an intellectual level is to publish unread articles in the journal stables of Elsevier, Sage, Springer, or Taylor & Francis. Journals held by such publishers are the pillar of legitimacy in the field of media studies and can be read without ever finding a concept. They are exercises in producing work that is largely indistinguishable from one article to the next. To put it very bluntly, the biggest threat to the future of media studies is the university itself.

The more recent work of Lev Manovich serves as one point of departure towards reinventing media pedagogy and concept production. Manovich frames his book *Software Takes Command* with the following question: "How does the shift to software-based production methods in the 1990s change our concepts of 'media'?"[9] Emphasizing the digital proliferation of visual material in contemporary life, Manovich foresees an accompanying crisis in critique. Where previously existing relations between critical approach and media form provided a sufficient architecture of intelligibility (think of psychoanalysis and film, for example), the abundance of digital culture today has yet to establish analytical models and relevant concepts with which to make sense of the hyperspeed of status-update culture. Teaching semiotics to your undergrads is not really going to help much when it comes to explaining the proliferation of YouTube videos, file-sharing sites, and social networking updates.

How to transform the culture of existing education institutions so that teaching is positioned as a generator of – rather than separate from – research presents a very particular set of obstacles. Part of the problem is that digital culture is so highly unstable. One of the key reasons the 19th century novel could obtain canonical status and help invent the critical field of literary studies was because generations of scholars were able to refer to and debate the minutiae of plots unfolding across individually produced pieces of content situated in the logic of an author's *oeuvre*. But this is not the case for digital culture, which is technically prone to collapsing code as programs are collectively rewritten and made obsolete with the latest update.

This technical characteristic manifests in social and cultural realms. For instance, there are no institutions beyond the idiosyncratic media art museums here and there that archive the internet's inherently

---

9   Lev Manovich, *Software Takes Command* (London: Bloomsbury Academic, 2013), 43.

unstable medium. Certainly there exist a scattering of online archives such as the Wayback Machine,[10] and indeed one could argue that competing platforms of search engines produce more than enough results for the networked masses.[11] But we are talking of something quite different than the algorithmic organization of hits and links. The production of concepts for analyzing digital media culture cannot exist exclusively online with all its social-technical ephemerality, economy of upgrades, and tendency toward high levels of distraction. On-the-ground institutional infrastructure that supports critical digital studies is required for sustained research agendas, the formulation of curricula, and support staff.

The next problem is that newer forms of teaching are assessed through managerial audit regimes obsessed with the self-referential and thoroughly time-wasting exercise of "quality assurance control." Those who do bring technology into the classroom often find and even willingly embrace a teaching process contained within institutional software systems – Blackboard, WebCT, and internal wikis are key culprits. As a consequence, teaching practices and content are not exposed beyond the borders of the institution and therefore not assessed by a world of peers at large.

A final problem of a practice-based approach is that it all too easily replicates the visual elements, hegemonic language, and software behaviors of its own tools. Critical concepts also do not just emerge out of practice. Teaching how to develop one's concepts should be distinguished from the techno-commercial skills to master and localize media. We see this so often in new media education: importation of hyped-up concepts from California and customizing them to make viable successful products. This sense of localization is indistinguishable from the publishing giants Sage or Taylor & Francis offering teachers the opportunity, if they pay, to remix a course reader from their database.

Could we engage Digital Humanities, with its separate IT staff who will deal with the technical details for clueless arts and humanities scholars? The concept of Digital Humanities might be useful for

---

10   HTTP://WWW.ARCHIVE.ORG/WEB/WEB.PHP.

11   The question of search engines served as one of the topics of investigation for the Institute of Network Cultures event, *Society of the Query*, Amsterdam, November 13–14, 2009, HTTP://NETWORKCULTURES.ORG.

retiring linguists but is deadly for the media studies environment. The humanities itself should try to overcome its distance from computer science and attack the sciences head on for its privileged position in terms of funding. Digital Humanities maintains to be working in collaboration with computer engineers but the problem is they retain a classic division of labor, which totally goes against transdisciplinary knowledge production. What we need is to move away from expert dependent high-end tools to broad-based undergraduate programs. We do not need massive budgets, but the Digital Humanities only work with high-end institutions and hide themselves behind the smoke screen of large datasets. The "digital" is revolutionizing the humanities itself and should not be downgraded as servitude to the IT staff and their authoritarian imaginations.

Mavericks, gurus, consultants, entrepreneurs, futurologists, and public intellectuals all trade in viral memes, but these are mostly bubble concepts that slide nicely across the PowerPoint presentation and are not connected in any material way to media research. This idiom of concept-lite derives from a lively and healthy publishing industry in business, management, and trade press – a post-war phenomenon that corresponds with the rise of consumer society and emergent post-industrialism. This extra-academic discourse is powerful precisely because thought becomes abstracted from material conditions. And at the level of visibility if not readability, it is very hard to ignore this literature – it can be found at every airport, yet it rarely appears on course reading lists.

Can we turn instead to innovative start-ups? Most of the work inside young companies involves the localization of existing realms of knowledge and commodity consumption – comparable to academics' endless reviewing, quoting, and copying of existing work under a banner of 'creativity' that suggests otherwise. Both fine-tune marketing and research paradigms around already existing knowledge. Similarly, in open source software development Linux is the basic platform and customization follows. The system itself is not transformed, but simply copy-pasted and then slightly altered. The rise of interconnected, integrated adaptation and localization of a handful of styles and ideas is preventing radical concept development. Where in this constellation described above that claims to be focused on critical, creative, innovative concepts does "research" fit in? Maybe research better positions itself in the tinkering, do-it-yourself corner of adaptation. The challenge is to organize research as forms of collective production and to modify existing research practices so a transformation can take place. This is what we must explore: "becoming a school."

So where does one turn to for concepts – the crystallization of ideas as frameworks for analysis? Media studies' future will only be guaranteed by the learning, teaching, and research of media taking place outside the academy, where it can turn instead to artist collectives and activist groups experimenting with and developing concepts for collaborative learning – concepts that can then be brought back into and applied in a classroom situation. The same could be said for many other disciplines – performance studies, art and design, etc. But somehow these disciplines have not become inherently dull in the way that media studies has. Given the choice, we would jump at the opportunity to work in design and architecture programs before landing a job in yet another media program with intellectually reluctant if not deeply cynical students. Why bother after all this? Because there is an incredible potential for collective experiences of self-organized modes of research development. Moreover, the profound connection between media and sociality is endlessly rich and holds the capacity to transform societies in substantial ways.

Media studies, *was nun*? What do we do, teach students to conduct social media marketing, build an iWhatever app, and play around with the latest data analytics tools? Is that all there is? Will it become an optional choice to include technical education in a media studies degree? Linux for freshmen? No. We need to discover (critical) concepts, which could be classified as mere luck but is a skill that can be trained nonetheless. One needs to develop an eye for them. All the rest is administration, calculating the inventory. In the fluid media environment concepts often hover on the surface but can only be recognized after intense intellectual expeditions into history, for example, mid-18th century Continental philosophy. Ideas then spring back to life in untimely contexts, where they become productive. Being informed about the latest hypes alone will not do the job. Only by making strange what appears common can one come up with concepts that capture new, surprising meanings and phenomena.

Organizing Concept Production

So what will happen if media studies finally moves from analyses of visual representations into a methods-based realm of post-visual culture? The challenge lies beyond, in the tactile, local, relational, haptic, invisible levels where thing to come reveal themselves. The work we collectively do as practitioners of media research resides in non-representational dimensions: mobility, miniaturization, and media integration into the urban realm. Should we rebrand ourselves, go with the flow, and rename our operations under larger umbrellas such as urban studies, design culture, or digital society? The sheer scale and proliferation of network media comprises complex environments, but it is not as though environments were any less complex than before. It's just that disciplines have trouble dealing adequately with informational and ontological complexity – to do so will require a transdisciplinary move. This is where the post-modern project comes to an end – the world refused to be deconstructed and took revenge.

Humanities academics have become more conservative and have regressed into intellectual restoration, with less concrete experience of how to organize things. Empirical simplicity has become the preferred mode of research in place of a reflexive and political understanding of processes and relations. Reflexive mediation these days is not limited to media *per se* but addresses the proliferation of borders, of affects, the multiplication of labor and subjectivity. The institutional and social landscape of work has become one defined by short-termism: portfolio careers and project driven jobs. There is no consistency other than the unstoppable waves of inconsistency. With informational integration the time of mediation is also the time of databases and information management systems. And there we find a place for political-economic research coupled with a practice that is focused, at least in our minds, on the production of open standards and open storage.

Media theory needs to unpack the subconscious terminologies that float around, such as friends, open, free, social, and community. As an example let's start with a critique of openness as a hegemonic rhetoric. There is a universal position across leftist culture that is anti-intellectual property. Anti-IP advocates are the noisiest in Old Europe, where one can still afford to be a cultural producer on the payroll of the state or the cushion of inherited wealth. But the media economy is moving beyond the IP system. The data economy today is based on the control of user profiles for recombination then repackaging for third-party clients desperate for affirmation of real consumption. Of course, the data miners such as Google and Facebook trade in fiction as much as the previous media economy, where advertising space and time was sold then broadcast to the masses.

In praise of conceptual thinking, we want to make a claim that collaborative methods immanent to media of communication – in short, the work of organization – can withstand time in an environment that sees platforms, software, and code languages coming and going at increasing speed. Within a decade no-one will know what an iPod was. Media studies should become more sustainable – it needs a projection into the future where the shelf-life of the concept is not as short as your next software update.

When programmers write a line of code they execute it within the sandbox to see if it works. Where is such a testing environment for media studies to be found? Science has its laboratories but where are ours? This situation can be seen in the pre-corporate and dull and formalistic way in which computer labs the world over are all the same: hostile settings of absolute sterility with security officers and cleaners as the only human element. What we need is more uncertainty, chaos, and untimeliness. Rise up outside the campus, open the doors and windows (equipment prices are no longer a large barrier to communication). Instead, we see students rolling in like sheep, turning on outdated locked-up hardware according to the rules and regulations stipulated by the IT police, and then going home. And these are supposed to be people educated for creative futures. It also has to be admitted that there is a profound reluctance on the part of the vast majority of media studies students to display the tiniest slither of intellectual curiosity. Let's break through these role models.

Concept production is intimately connected to the challenge of method, of how we operate in specific situations (whether online or offline). In the global economy, for instance, the researcher can have a new role entirely in the collective analysis of workplace settings that are dependent on the use of digital media and software applications in daily business. Think of the global logistics industries and their management of labor and supply chains. There is enormous scope for the birth of a new field of software studies to analyze the effects of logistics software in the production of organizational systems and subjectivity.

Instead of promoting the informality of the existing social networking sites, we propose to experiment. We champion orgnets here not as an identity or brand. Similar names are plentiful and there should be more, starting with barcamps, temp media labs, unconferencing, summer schools, book sprints, and master classes. We need to theorize these unruly practices in education. Just think if Christopher Kelty's "recursive politics" concept could be inserted into the aesthetics agenda of software studies.[12] Or sonic studies with radio as a technical media of writing, and DJ Spooky as prime theorist.[13] In order to get initiatives up and running we have to glance away for a moment from the busy screen: Oublier le Pop, Lady Gaga, or *South Park* as organized networks? Forget it. There is no match or correspondence. Critique is not a mood or sentiment, but a plan to organize things in a different manner.

---

12 See Christopher M. Kelty, *Two Bits: The Cultural Significance of Free Software* (Durham: Duke University Press, 2008).

13 See Paul D. Miller (aka DJ Spooky), *Rhythm Science* (Cambridge, Mass.: MIT Press, 2004). See also the highly inventive work of Australian sonic theorist and avant-pop star, Philip Brophy, HTTP://WWW.PHILIPBROPHY.COM.

6.

SERIALITY, PROTOCOLS, STANDARDS

Recommended music track by the Ramones to play
while reading:
HTTP://WWW.YOUTUBE.COM/WATCH?V=Z6XAE9JSQXU

TODAY YOUR CODE, TOMORROW THE WORLD. "WHOEVER SETS THE standard has the power." Strangely enough, this view has few disciples. If we talk about power, and dare to think that we can take over and be in charge, we rarely take Voltaire's advice to focus all our attention on victory and instead indulge ourselves in self-criticism over how time and again we fail. Mention the word power and we will almost intuitively think of the political class and our revulsion for this profession. We prefer to believe media-savvy opinion makers control the political agenda. It is tempting to think that content, and not form, determines our lives. Those of us who publicly discuss protocols are easily dismissed as cynical techno-determinists or boring bureaucrats. The standard height of a computer table is 72cm. But who gets bothered about that? Isn't it the quality of the work that comes out of the computer on that very table which counts? An easy-on-the-eye font for a novel is nice enough, but what really counts is the writer's gift to entertain us.

For many years, philosophers have been casting doubt on the common identification with meaning and signification as the primary human response mechanisms to the world. If we wish to understand anything about how our complex technical society is made up, we must pay attention to the underlying structures that surround us, from industry norms to building regulations, software icons and internet protocols. Yet our ordinary understanding of the world resists this very idea. If we call for another society, with more equality and style, it is not enough to think differently; the very framework of that thinking must be negated and overturned. Or implode, vaporize, fade away (if you are in Baudrillard mode). A "true" revolution in today's technological society is not one where politicians are replaced but all the very standards and protocols of the system as such are overthrown, or at least put into question. Media experts are already aware of this. If you want to make a lasting contribution that makes a substantial difference you will have to design the standards for communication. It is not enough to unleash a Twitter revolution: you have to develop – and own – the next Twitter platform yourself. This is the politics of the standard: those who are able to determine the outline of the form determine like no other tomorrow's world.

"Protocol Now!" We cannot deal with the unbearable truth that techno-determinism confronts us with. Reducing the world to rules that rule us shuts down the imagination and turns the otherwise docile and routine-minded Western subject into a rebel. Protocols, so we fear, cannot be questioned and are looked down upon as religious rituals that have lost their original social context. What remains are empty, meaningless gestures. Why is it that only a few attach any belief to the essence that hides behind our technical infrastructures? Guided by the real-time attention economy we are so easily distracted, moving from one surface to the next. The very idea that, when it actually comes down to it, a closed company of technocrats decides our window on the world should be cause for concern. It is not supposed to be the HD camera or the animation program that makes a film good or bad but the creative skills of the filmmaker to tell the story in such a way that we immediately forget the technical details. At least this is how we are repeatedly inclined to think in this Age of the Amateur.

Who really understands the degree to which the browser decides what we get to see on the internet?[1] Who will finally map the influence that giants like Google, Apple, Facebook, and Microsoft have on our visual culture? Marshall McLuhan's sixties' statement that "the medium is the message" remains a misunderstood speculation, which has proved not untrue but rather irritating in its banality. The attention paid in the media to background standards and protocols is minimal. Instead, we gaze starry eyed at the whirlwind lives of celebrities and the micro-opinions of presenters, bloggers, and commentators. It is this sort of interference that reassures us. But when will the discomfort with the artist as an eye candy maker actually emerge?

In his 2004 book *Protocol* New York media theorist Alexander Galloway introduced a critical theory and humanities reading of the protocol concept. The book addressed a young audience of geeks, artists, scholars, and activists that were not primarily interested in legal issues, or the bureaucratic side of technology for that matter.[2] Even though only a part of this edited PhD deals with the topic itself, the book quickly became popular – the main reason for this being that the author asked the often-heard question of how control can exist in distributed, decentralized environments. This interest coincided with the popularity of the "network of networks" image of the internet and its self-correcting automated processes and auto-poetic machines that never stop producing and are so efficient exactly because there is no top-down interference of a Big Brother. The trick here is to read the liberating prose of the engineers and IT management gurus against the grain: the autonomous nodes are the new manifestation of power.

---

1 See Konrad Becker and Felix Stalder (eds), *Deep Search: The Politics of Search Beyond Google* (Innsbruck: Studien Verlag, 2009).

2 Alexander R. Galloway, *Protocol: How Control Exists after Decentralization* (Cambridge, Mass.: MIT Press, 2004). A related title, often read in contrast to Galloway, is Wendy Chun, *Control and Freedom: Power and Paranoia in the Age of Fiber Optics* (Cambridge, Mass., MIT Press, 2006) – books which, as Amazon.com recommends, are "frequently bought together." An earlier, sociological approach to the global politics of internet governance and domain names in particular is given by Milton Mueller in *Ruling the Root: Internet Governance and the Taming of Cyberspace* (Cambridge, Mass., MIT Press, 2002).

For Galloway "protocol refers to the technology of organization and control operating in distributed networks."[3] Ever since Galloway made the rounds in new media theory and art circles there has been increased talk about protocological power inside network cultures. The term is used to open up a dialogue over forms of power where we see no power. Protocol is introduced not so much as a fixed and static set of rules but rather to emphasize processual flows. In this view protocols create the shape of channels and define how information and data are embedded in social-technical systems.

The Agony of Power

The protocolization of society is part of wider transition from domination to hegemony, as Jean Baudrillard explains in *The Agony of Power*.[4] In the old situation domination could only be reversed from the outside whereas under the current regime hegemony can only be inverted from the inside. "Hegemony works through general masquerade, it relies on the excessive use of every sign and obscenity, the way it mocks its own values, and challenges the rest of the world by its cynicism."[5] Protocols are the invisible servants of the new Soft Power. In order to resist and overcome the ruling protocols, or should we say the protocological regime, Baudrillard states that we cannot go back to the negative and should instead emerge into the "vertigo of the denial and artifice." Leaving behind the theatrics of refusal we can start exploring "total ambivalence" as a strategy for overcoming the protocological determination of resistance.

If we look at contemporary internet culture we can quickly notice that the "participatory culture" of Web 2.0 and social media are not run by artistic windbags but average IT engineers whose job it is to implement software according to the given protocols. The internet is part of popular culture that focuses its values on "the crowds" and is not the least bit interested in early 20th century activities of the avant-garde. The "cool" image of the creative IT clusters should not confuse us. The unprecedented exercise of power by these start-ups is no longer part of a conspiracy. What is being executed here are MBA

---

3   Alexander Galloway, "Protocol," *Theory, Culture & Society* 23.2-3 (2006): 317.

4   Jean Baudrillard, *The Agony of Power*, trans. Ames Hoges (Los Angeles: Semiotext(e), 2010).

5   Ibid., 5.

scenarios in which venture capitalists have the final say. Control 2.0 is no longer centered around individuals and their ideologies; it is decentralized and machine-driven, some would even call it a topological design of continual variation. This makes it more difficult to identify who really calls the shots. It seems that power is no longer in the hands of people, but manifests itself in software-generated social relationships, surveillance cameras, and invisible microchips. So what can we do with this insight? Does it make us depressed because we do not know where to start, or rather joyful as we can simply hack into their leaking wikis and delete data, without harming humans, in case we want to take over? Or would we rather like to dismantle power as such? In that case, would it be possible to sabotage the very principle of "protocol" itself? Is it sufficient to openly display the dysfunctionality of the system or should we expect to also come up with a blueprint with viable alternatives before we attack?

"Power to the Protocol." Working out who defines and manages the technological standards has become a new method of power analysis. Protocol once referred to a tape with verification and date stuck to a papyrus roll. Now, protocol is promoted to a decisive collection of ambivalent and implicit rules on which today's complex societies revolve. The protocol meme has turned cool and indicates that you understand how the shop is run after control. How can we get a grip on the invisible techno-class that prescribes these rules? Is it sufficient to urge participation? Demonstrating the undemocratic character of the closed consultation is one thing, but are alternative models available? Is it sufficient to discover the holes and bugs in protocols? What do we do with our acquired insight into the architecture of search engines, mobile telephone aesthetics, and network cultures? It is one thing to become aware of the omnipresence of protocols at work. But what to do with all these insights? These are also questions about the politics of knowledge production and the production of new subjectivities.

Complex problems such as human rights violations, climate change, border disputes, migration control, labor management, and the informatization of knowledge hold the capacity to produce trans-institutional relations that move across geocultural scales, and this often results in conflicts around the status of knowledge and legitimacy of expression. A key reason for such conflicts has to do with the spatio-temporal dynamics special to sites – both institutional and non-institutional – of knowledge production. Depending on the geocultural scale of distribution and temporality of production, knowledge will be coded with specific social-technical protocols that give rise to the problem of translation across the milieu of knowledge. This is not a question of some kind of impasse in the form of disciplinary borders, but a conflict that is protocological.

The most often used example of here is the internet and its TCP/IP protocol, but in their book *The Exploit: A Theory of Networks* Alexander Galloway and Eugene Thacker also include DNA and bio-politics in their analysis of protocological control.[6] Another example could be the global logistics industry whose primary task it is to manage the movement of people and things in the interests of communication, transport, and economic efficiencies. One of the key ways in which logistics undertakes such work is through the application of technologies of measure, the database and spreadsheet being two of the most common instruments of managerial practice. In the case of cognitive labor, the political-economic architecture of intellectual property regimes has prevailed as the definitive instrument of regulation and served as the standard upon which the productivity of intellectual labor is understood. This is especially the case within the sciences and increasingly within the creative industries, which in Australia and the UK have replaced arts and humanities faculties at certain universities.

Materiality of Informatized Life

There are, however, emergent technologies of both labor management and economic generation that mark a substantial departure from the rapidly fading power of IPRs, which is predicated on state systems enforcing the WTO's TRIPS Agreement – something that doesn't function terribly well in places like China with its superb economies of piracy or in many countries in Africa where generic drugs are subtracting profits from the pharmaceutical industry and its patent economy.[7] Intellectual property rights are no longer the site of real

---

6   Alexander R. Galloway and Eugene Thacker, *The Exploit: A Theory of Networks* (Minneapolis: University of Minnesota Press, 2007).

7   See the fascinating work of Melinda Cooper, who has been studying the economy and geopolitics of clinical labor trials within the pharmaceutical industries – the rise of which can partly be seen as a way of offsetting profits lost from the diminishing returns availed through IPRs as a result of the increasing availability of generic drugs, which in turn can be understood as a sort of pirate

struggle for informational labor, although they continue to play a determining role in academic research and publishing when connected to systems of measure, such as global university and journal rankings, "quality assurance" audits of "teaching performance," numbers of international students, and so forth. In the age of cognitive capitalism, new sites of struggle are emerging around standards and protocols associated with information mobility and population management in the logistics industries. Key, here, is the return of materiality to computational and informatized life.

Like protocols, standards are everywhere. Their capacity to interlock with one another and adapt to change over time and circumstance are key to their power as non-state agents of governance in culture, society, and the economy.[8] Standards require a combination of consensus and institutional inter-connection (or hegemony) in order to be implemented through the rule of protocols. In this way, one can speak of environmental standards, health and safety standards, building standards, computational standards, and manufacturing standards whose inter-institutional or technical status is made possible through the work of protocols. The capacity for standards to hold traction depends upon protocological control, which is a governing system whose technics of organization shape how value is extracted and divorced from those engaged in variational modes of production.

But there can also be standards for protocols. As mentioned above, the TCP/IP model for internet communications is a protocol that has become a technical standard for internet based communications. Christopher Kelty notes the following on the relation between protocols, implementation and standards for computational processes: "The distinction between a protocol, an implementation and a standard is important: *Protocols* are descriptions of the precise terms by which two computers can communicate (i.e., a dictionary and a handbook for communicating). An *implementation* is the creation of

---

economy that even intersects with aspects of open source cultures. Melinda Cooper, 'Experimental Labour-Offshoring Clinical Trials to China', *EASTS East Asian Science, Technology and Society: An International Journal* 2.1 (March 2008): 73–92.

8   See Martha Lampland and Susan Leigh Star (eds), *Standards and their Stories: How Quantifying, Classifying and Formalizing Practices Shape Everyday Life* (Ithaca: Cornell University Press, 2009).

software that uses a protocol (i.e., actually does the communicating; thus two implementations using the same protocol should be able to share data). A *standard* defines which protocol should be used by which computers, for what purposes. It may or may not define the protocol, but will set limits on changes to that protocol."[9]

A curious tension emerges here between the idea of protocols as new systems of control and standards as holding the capacity to limit that control. Without the formality of "cold" standards, the "warm" and implicit, indirect power of protocols is severely diminished. Standards, in other words, are the key site of the politics of adoption. Herein lies the political potential for the Revenge of the Masses. Social media serve as a good example: their strength is only as good as their capacity to maintain a hegemony of users. Think of what happened to MySpace once Facebook and Twitter took off in 2006 as the preferred social media apps. Having paid a crazy $580 million in 2005, Murdoch's News Corp dumped its toxic asset for a paltry $35 million in June 2011, contributing to a 22% fall in profits in the quarter to June. The effective collapse of MySpace signals that while masses might not build standards at a technical level, they certainly hold a powerful shaping effect that determines whether a standard becomes adopted or not.

Invent Your Own Standards

The next step is to decide, if not invent, our own standards and protocols in the world of social media. Let's move beyond the dependency on Google, Facebook, and Twitter for political organization. Just a few years ago we saw what happened during the London student protests against government funding cuts to education – numerous activists groups were deactivated and user accounts suspended by Facebook administrators without advance warning. There are distributed social media alternatives out there, and activists do need to be aware of the political implications of assuming communication protocols established by corporate media. Once the protocological layer of rules is established the political work of building affiliations around standards begins. How to organize the distributed and often conflicting interests

---

9   Christopher M. Kelty, *Two Bits: The Cultural Significance of Free Software* (Durham: Duke University Press, 2008), 330n28. Available at: HTTP://TWOBITS.NET.

of users around the work of creating robust standards is a key challenge for the next decade in which we will see major battles between large scale monopolies, increased state control, and decentralized, networked initiatives.

When our research projects and knowledge production become fragmentary exercises that cleave open intermittent gaps in time both orchestrated and unforeseen, we require a catalogue of standards to help maximize the research outputs and amplify the experience of intellectual and social intensity. Various models have been tested in recent years, including the enormously popular speed-dating phenomenon of PechaKucha, which celebrates intellectual emptiness with 20 slides shown in 20 second intervals in design, fashion, and architecture circles. At the high end of the tech-design scale, the TED Talks aim to populate the world with Silicon Valley inspiration. Critical research on network societies and information economies also needs to generate its own standards.

The design concept and practice of "seriality" offers one technique and strategy of "total ambivalence" with respect to organizing networks in ways that establish autonomous standards in this proto-cological society of control. As a term, seriality suggests some kind of correspondence with standardization as critiqued by Adorno and Horkheimer in their essay on the "culture industry."[10] The standardized distribution and production techniques in the film industry were seen by Adorno and Horkheimer as an industrial rationale to address the organization and management of consumer needs and the desire for uniformity – or what in the education industry today is referred to as "quality assurance" – in the experience of cultural consumption. But we need to distinguish seriality from the Frankfurt School critique of standardization and industrial production, which is typically seen as synonymous with the Fordist assembly line and the production of undifferentiated docile subjects.

Designing Seriality

Seriality is a line to future possibility. McLuhan knew this well with his concept of probes. The line for us is not about orthodoxy

---

10 Theodor Adorno and Max Horkheimer, "The Culture Industry: Enlightenment as Mass Deception," in *Dialectic of Enlightenment*, trans. John Cumming (London: Verso, 1979), 120–67.

or conforming to a dominant protocological rule. When we deploy the term we do not invoke the well-known phrase "toe the line...." Instead, the line is a connecting device that facilitates production across otherwise disaggregated networks that are so often compelled to start from scratch when they begin a new venture or undertaking. The line enables the production of standards from a core, which is not equivalent to network metaphors of a hub with nodes or parent company with subsidiary corporations.

A design core is a distributed accumulation of practices, skills, lessons, capacities, connections, concepts, strategies, and tactics that build on collective experience over time. The design challenge is to bring these variegated elements into relation in such a way that they can be communicated and develop into standards. This is where organization enters in order for seriality to find its line of continual differentiation. Computer engineers are often tasked with this challenge but typically they step in to a project and tinker with the code and then leave for the next job. For some weird reason they are not seen often enough as crucial to the continuum of network cultures.

Serial design is not system design or industrial design. *System design* defines the parameters, standards, and modularity necessary for product development. *Industrial design* is primarily about scale, inasmuch as aesthetics and usability are integrated in order to maximize extension. *Serial design*, on the other hand, also brings aesthetics and usability into a constituent relation, but it is not so much about scale itself. We do not need to expand our networks infinitely or extend them across space. We are not talking even about global networks. Instead, it is the problem of time that seriality engages in the time-starved universe of data economies and cultures. We are all so pressed for time that we download the next widget in a hopeless gesture toward time-saving devices.

Hacklabs, barcamps, unconferencing, book sprints, mobile research platforms – these are all formats that through the work of seriality have become standards for network cultures. Combining on and offline dimensions, they are designed to maximize techno-social intensity and collectively develop products and accumulate experiences in a delimited period of time. Their hit-and-run quality might give the appearance of some kind of spontaneous flash-mob style raid, but in fact they are carefully planned weeks, months, and sometimes years in advance. Despite the extended planning duration and intensive meeting space of these formats, they are notable for the way in which they occupy the vanguard of knowledge production. Only five or so years down the track do we see the concepts, models, and phenomenon of these formats discussed in academic journals, by which time they have been drained of all life. A key reason for this has to do with the flexibility these formats retain in terms of building relations and taking off into unexpected directions as a result of the unruliness of collective desires. We might see this as the anti-protocological element of network cultures.

Nevertheless, the problem of sustainability still plagues the work of organization across network cultures. Certainly, the practice of seriality goes a long way to addressing this problem. But we need to advance the discussion and practice of seriality by connecting it to the collective development of standards. Only then will we begin to formalize seriality as a distributive practice that can sustain network cultures as new institutional forms. Once that happens, the conflict surrounding the hegemony of protocological regimes will come into full swing.

The hard reality once preached by the historic avant-garde is still valid, no matter how disastrous the implementation of utopian programs may have been. There is an increasing number of artists who have the ambition to sketch the framework of society. They design new rules and do not simply produce cool design. What we must look for are the contemporary variants of Google. This media giant, with internet pioneer and ICANN domain name boss Vint Cerf (jointly) ruling in the background, is a perfect example of how economic, political, and cultural power can be built up using technical laws (algorithms). We can do that as well. The connected multitudes, now in their billions, have reached the end of a long period in which the workings of power first had to be understood and subsequently dismantled. What we are designing now are new spaces of action. Before we concentrate on open standards for living, work, and play, we should open a public debate about this matter. Can the loose networks of today organize themselves in such a way that they set the rules for tomorrow's communication? Yes We Can: Set the Standard.

7.

FINANCING NETWORKS

ORGANIZED NETWORKS HAVE TO BE CONCERNED WITH THEIR OWN sustainability. Networks are not hypes. We've past the nineties and that potlatch era will not return. Networks may look temporary but are here to stay, despite their constant transformations. Individual cells might die off sooner rather than later but there is a Will to Contextualize that is hard to suppress. Links may be dead at some point but that's not the end of the data itself. Nonetheless networks are extremely fragile. This may all sound obvious, but let's not forget that pragmatism is built upon the passions, joys, and thrills of invention. Something will be invented to bridge time and this something we might call the organized network. Time has come for cautious planning. There is a self-destructive tendency of networks faced with the challenge of organization. Organized networks have to feel confident about defining their value systems in ways meaningful and relevant to the internal operations of their social-technical complex. That's actually not so difficult. The danger is ghettoization. The trick is to work out a collaborative value system able to deal with issues such as funding, internal power plays, and the demand for "accountability" and "transparency" as they scale up their operations.

So let's get monetary. Organized networks first and foremost have to keep their virtual house in order. It is of strategic importance to use a non-profit provider (ISP) and have backups made, or even run a mirror in another country. Also, it is wise not to make use of commercial services such as Yahoo!Groups, Hotmail, Geocities, or Google, as they are unreliable and suffer from regular security breaches. Be aware of costs for the domain names, e-mail addresses, storage, and bandwidth, even if they are relatively small. Often conflicts arise because passwords and ownership of the domain name are in the hands of one person that is leaving the group in a conflict situation. This can literally mean the end of the project.

Networks are never 100 per cent virtual and always connect at some point with the monetary economy, which, in case we've forgotten, is in so many ways a material culture. This is where the story of organized networks start. Perhaps incorporation is necessary. If you do not want to bother the network with legal matters, keep in mind what the costs of not going there will be. Funding for online activities, meetings, editorial work, coding, design, research, or publications can of course be channeled through allied institutions. Remember that the more online activities you unfold, the more likely it is that you will have to pay for a network administrator. The inward looking free software world only uses its paradise-like voluntary work rules for its own coding projects. Cultural, artistic, and activist projects do not fall under this category, no matter how politically correct they might be. The same counts for content editors and web designers. Ideally, online projects are high on communitarian spirits and are able to access the necessary skills. But the further we leave behind the moment of initiation, the more likely it will be that work will have to be paid. Organized networks have to face this economic reality or find themselves marginalized, no matter how advanced their dialogues and network use might be. Talk about the rise of "immaterial labor" and "precarious work" is useful, but could run out of steam as it remains incapable of making the jump from speculative reflection to a political agenda that will outline how networks can be funded over time.

Organized networks are always going to face great difficulty in raising financial resources through the traditional monetary system. It is not easy to attract funding from any of the traditional sectors of government, private philanthropy, or business. Alternatives need to be created. Arguably, the greatest asset of organized networks consists of what they do: exchanging information and conducting debates on mailing lists; running public education programs and archiving education resources; open publishing of magazines, journals, and books; organizing workshops, meetings, exhibitions, and conferences; providing an infrastructure that lends itself to rapid connections and collaborations amongst participants and potential partners; hosting individual web sites, wikis, blogs, etc.

All of these activities can be understood as media of communication and exchange. This quality lends itself to translation into what Bernard Lietaer – co-designer of the Euro and researcher of complementary currencies – defines as currency in its multiple uses and forms: "an agreement within a community to use something as a medium of exchange."[1] Lietaer says there are over 4000 forms of complementary currencies worldwide, from the customer loyalty systems of frequent flyer memberships to community development currencies in Bali. The LETS system is perhaps one of the better known alternative forms of complementary currency for those in the West. More recently, the hype around cryptocurrencies and Bitcoin has captured the imagination of many with all the excitement around decentralized blockchain architectures. But let's never forget that decentralized systems are really only distributed within and across centralized hubs of communication and storage.

In Japan, credit-for-care tickets can be accumulated for services not supported by the national health insurance system. Credits can then be used to pay for university tuition fees, or they may be transferred to another family member who is in need of domestic assistance. Lietaer makes reference to a survey in which elderly people in Japan preferred care services paid for with "fureai kippu" (caring relationship tickets) over services paid for in yen. Such a form of affective labor addresses many of the problems and difficulties faced by ageing populations.

The primary difference between conventional and complementary currencies rests on the different regimes of value inscribed upon the mode of labor and the logic of exchange. Lietaer: "Conventional

---

1   Ravi Dykema and Bernard Lietaer, "Complementary Currencies for Social Change: An Interview with Bernard Lietaer," *Nexus: Colorado's Holistic Journal* (July–August, 2003), HTTPS://UAZU.NET/MONEY/LIETAER.HTML.

currencies are built to create competition, and complementary currencies are built to create cooperation and community...." The tension between multiple currency systems constitutes a form of mixed economies, and mitigates any tendency to get washed away by the euphoria of feel-good complementarity.

If there is a decision to be made, and an enemy to be singled out, it's the techno-libertarian religion of the "free." It's high time to openly attack the cynical logic of do-good venture capitalists that preach giving away content for no money while making millions of dollars in the back room with software, hardware, and telco-infrastructure, which the masses of amateur idiots need in order to give and take for free. Organized networks are wary of the gurus on high consultancy fees who "inspire" others that they should make a living out of selling t-shirts: "You poor bugger, fool around with your funky free content, while we make the money with the requirements." It is time to unveil this logic and publicly resist it. Knowing is not enough.

The key point of networks is not so much their form of organization but the fact that the business model has been on the agenda. The networked organization, however, is setting the terms for entry into economic sustainability. Whereas the precursors to the organized network – lists, collaborative blogs, alternative media, etc. – are used to being on the vanguard of inquiry and practice, at the same time there is an undeniable distrust towards the networked organizations. For too long the ghetto of list cultures has resulted in a self-affirmation that is now a major obstacle to the possibility to scalability. What is required for the organized network to scale up? A transparency of formalization and shift in the division of labor? It is well known that formal networked organizations are the darlings of funding bodies, whereas real existing networks miss out because they fail to undertake the proper lobby work and cannot adequately represent themselves. It is ironic that it is exactly the global nature of networks that makes it next to impossible to fund them. There are no global funds for global networks – despite all the nineties rhetoric.

Free Culture Costs Money

There is no universal recommendation or model for practitioners in the creative industries. Creative practice consists of what Spivak terms "irreducible idiomatics" of expression. One size does not fit all, in other words. You wouldn't spot this if you limited your reading list to government policy, however. A universal definition does exist within this realm: creative industries consists of "the generation and exploitation of intellectual property." In all seriousness, how many creative practitioners would call themselves producers, let alone financial beneficiaries, of intellectual property? Most probably don't even know what IP means. We must redefine creative industries outside of IP generation. This is the dead-end of policy. When understood as "the generation and exploitation of intellectual property," creative industries registers the "banal evil" of policy mentalities, and assumes people only create to produce economic value. There needs to be a balance between alternative business models and the freedom to commit senseless acts of creativity. The tension between these two constituent realities is what needs to be investigated.

There are also severe limits to the "open cultures" model that stems from libertarian and open sources cults. The free culture model is essentially a North American libertarian view of the world in its own image. European activists are quick to reproduce this and, in avoiding the question of money trails and connections, also avoid engaging key actors and issues that comprise the political of network society and data economies. Taken as a Will to Conformity, free culture serves as a political retreat that parades as radical self-affirmation.

Touching the auto-erotic drive to create without purpose, collaboration, and the anarchistic rubric of mutual aid escapes these endless chains of re-appropriation. But they lack suspicion of instrumental intentionalism. These issues were the topic of a thread on the MyCreativity mailing list in 2007 following a posting of a report in *Spiegel* magazine ranking Berlin as the number one "creative class" city based on classic Richard Florida indicators: in this case, what has been termed the "3T's" – Talent, Technology, and Tolerance.[2] The seductive power of such indicators inspires the proliferation of hype-economics, transporting Berlin from a "poor but sexy" city to an economic nirvana populated by cool creative types. But the problem with such index obsession is that it functions through circumscription and the exclusion of a broader range of economic indicators that contradict such scenarios. In its 2007 city-ranking review, *Wirtschaftswoche* (Economic Weekly) undertook a comparison of 50

---

2   See "Berlin Tops Germany for 'Creative Class'," *Spiegel*, October 10, 2007, HTTP://WWW.SPIEGEL.DE/INTERNATIONAL/BUSINESS/0,1518,510609,00.HTML.

German cities according to employment, income, productivity, and debt. Berlin came in at number 48.[3] What does this say about Berlin's 3T's of creative economy? You can only conclude that the correspondence between indices and material realities are best left for policy fictions – despite all the groovy building sites along the Spree river. Indicators never end. Any number of permutations is possible. But government policy-makers and corporate beneficiaries are rarely keen to promote a negative future-present. It is precisely these sorts of reasons that necessitate the counter-research advocated by MyCreativity. Media theorist Matteo Pasquinelli proposes an analysis based on a Negative Index:

> Actually what I see is the risk of a "Barcelonisation" of Berlin, named after the touristic turn of Barcelona that transformed its cultural and political heritage into a theme-park for a young rich global class. The legendary Berlin underground is under the process of a slow gentrification (you can gentrify even "intangible assets"). "Barcelonization" means a parasitic economy and not a productive one, an economy based on real estate speculation and passive exploitation of natural resources (sun and good food for example): is such an economy "creative," productive? Is that a model we can apply to Berlin? Still the most affordable capital of Europe (especially East Berlin), some think that the speculative mentality will never conquer Berliners as they are used to [cheap] rent and live on social housing. Will Berlin's cultural industries develop a "parasitic" economy based on speculation, local consumption, and imported capitals or a productive economy based on production of knowledge/cultural and exportation of immaterial products? And what will be the impact of the Media Spree speculation (www.mediaspree.de) on the East Berlin cultural ecosystem?[4]

---

3   See "Die erfolgreichsten Städte Deutschlands," *WirtschaftsWoche*, 2007, HTTP://ONWIRTSCHAFT.T-ONLINE.DE/C/83/96/99/8396998,PT=SELF,SI=1.HTML.

4   Matteo Pasquinelli, "Re: [My-ci] Berlin Tops Germany for 'Creative Class'," posting to mycreativity mailing list, October 15, 2007, HTTP://IDASH.ORG/

An army of sociologists and cultural researchers is slowly assembling around questions such as these. The creative industries meme dominates research funding calls in the humanities, after all (particularly in Europe in more recent years). But don't expect to read the results too easily – they come at a cost as well, with the vast majority of academics happily transferring their results of state-funded research into commercial publishing houses that charge crazy fees for access to their journals. Organization and management researcher Steffen Böhm responded in reflexive style to Pasquinelli: "I think it would be good to understand the process of how activists (like people on this list) and the communicational economy that this list is part of is the very vehicle that helps to create a speculative bubble around certain issues/places/things/symbols. In other words, how is it that critics of the system become the 'driver' of the restructuring and transformation of that very system, enabling it to capture new forms of re-production?"[5]

Böhm attributes an influential power to critics and their capacity to shape the creative economies that is debatable. It is less the case of critics becoming drivers of bubble economies as it is the rise of cheap airlines determining markets for easy consumption. But he is correct to observe that critics and activists are agents within what he elegantly terms the "communicational economy" of creative industries. How, though, to maximize this critical potential in ways that do have concrete impacts on the development of creative industries research and policy formation? What other models can there be for concept generation that goes beyond the easyJet mobility of the commuting class, boozing masses, and conference circuits?

Efforts at transdisciplinary research are important here. The collective input of artists, designers, academics, policy-makers, and activists is crucial. General concept development and detailed case studies are not a contradiction. Empirics interpenetrates concepts, and vice-versa. Of course we can't take such research collaboration for granted. Not only are there considerable disciplinary and paradigmatic differences to negotiate, but there are also the banal practicalities of assembling

---

MAILMAN/LISTINFO/MY-CI.

5   Steffen Böhm, "Re: [My-ci] Correction – Berlin Tops Germany for 'Creative Class'," posting to mycreativity mailing list, October 18, 2007, HTTP://IDASH. ORG/MAILMAN/LISTINFO/MY-CI.

people in a particular place in order to meet. Not everything can happen online. Beyond mailing lists and collaborative blogs, perhaps networked academies and distributed think tanks are models for accommodating future critical research on cultural economies.

Angel Investors

The Libertarian Ideology hides its own mechanisms of making money. Libertarian open source movements are no different at the level of structure, organization, and financing from the monopoly of corporations involved in video game production. Tactically they focus on the right to remix, the basis of all creativity. Sure, this is nice. It goes back to the idea that all culture is distilled from a basic, common source. Organized networks wish to undertake projects, and to do this requires resources and financing beyond simply a capacity to mix code. In this sense, there is a parallel here to organized crime, whose aim is to redistribute stolen resources and property.

Organized crime is involved in translation. In terms of what networks are and ought to be, this element is consciously excluded in the software architecture and beyond. The repurposing and redirecting of financial resources appropriated by organized criminal networks is precisely what enables them to proliferate. Organized networks have a lot to learn from the creativity of criminal industries if they wish to address the problem of sustainability. So here's your "get out of jail free" card: criminal networks can be understood as an equivalent resource to the "presence of organized networks of individual angel investors."[6]

Since organized networks are seemingly in a condition of perpetual exclusion from conventional, institutional modes of financing, then there is really only one option left: to leave the network or, alternatively, to understand the logic of crime. There isn't much to obtain from the open source gurus. At least they have not totally captured the attention of so-called internet culture and research. Instead, they have migrated over to traditional cultural institutions, which now consider open source as the primary model. This will be an interesting experiment to observe, since the open source model goes against the border

---

6   Edward Lowe Foundation (ELF), "Building Entrepreneurial Communities," 2002, HTTPS://WWW.NATCAPSOLUTIONS.ORG/LASER/LASER_BUILDING-ENTREPRENEURIAL-COMMUNITIES.PDF.

controls of the traditional institution. Whether such institutions are able to fully embrace the logic of open distribution and retain both their brand and funding capacity remains to be seen.

Given that the organized network has no financial basis for its activities, why, then is accountability an issue here? This, of course, relates back to the question of transparency, governance, and control, and thus the structural dynamics of networks. This is a matter of making visible the capacities of the network to undergo transformation precisely due to the way in which accountability reveals limits. What does accountability mean outside the framework of representation? What does representation mean within a post-representative political system? How does it work?

8.

UNDERSTANDING CARTOPOLITICS

THERE HAS LONG BEEN A RELATION BETWEEN AESTHETICS, CULTURE, and systems of knowledge. The rise of the network society has not been exempt from this, with a vast range of efforts seeking to represent any manner of networks: activist movements, migration patterns, corporate monopolies, stock-market flows, neurological systems, to say nothing of the proliferation of social network cultures. Some of these representations can be very sophisticated analytical tools and aesthetically fascinating. The French collective Bureau d'études comes to mind, but also Lev Manovich's cultural analytics. We get a mighty, and humble, feeling of planetary overview. Ask the kosmonauts. Yet there is a danger emerging: politics runs the risk of being displaced by aesthetics. Walter Benjamin already warned of this in the 1930s and the aestheticization of politics has been a traumatic signal of social decline ever since. This is the problem of representation as such. Whereas visualization tools make it easy to create interactive maps, the question we ask here comes from inside the (visualized) networks themselves: it might be handy for researchers to be able to navigate through these data sets, but what's the point of this for the actors themselves? The empowering aspects, outside the safe walls of universities and NGOs, are often unclear. Do the conceptual insights of a cartographic overview lead to critical practices, as its promoters claim? What can you do if you are networked yet resist being mapped? We need to know more about the ever-present Will to Visualize.

Is it useful to distinguish between networks as living, ever-changing entities and dead information? Maybe there is nothing wrong with the visual porn of slick database visualizations if they make it easier for us to search and browse through millions of files, entries, pictures, or tags. Just as it is hard to imagine a world without search engines, is this also the case for data visualizations? These days networks are vital forms of social life. And in turn, they shape the social. People will almost intuitively organize themselves into networks, meaning that they have a commitment to some and "weak ties" with many of the possible members within the network. A network can grow quickly and have thousands involved, but can also remain very small. They can happen overnight and disappear again the next week. Having mixed-up private friends and work contacts helps campaigners to reach millions, but also gives operators of social media sites like Facebook and Twitter (and affiliated secret services) an unprecedented insight into our lives. The question of the visibility of such processes is accompanied by another question – who does this visibility serve? Do we eagerly display our network of sociality for all on Facebook in order to be subject to data mining economies? Does it matter that your future boss, or perhaps even the cops, knows about your dirty weekend, which books you read, the music you prefer, or party you voted for? Or is the representation of networked life really something more self-referential, indexing the parametric design of visualization software, and therefore prompting the hypothesis that perhaps indeed networks are always invisible, beyond representation?

Let's take the case of WikiLeaks. There is an aestheticization of WikiLeaks in the move from informal activist networks to the realm of Big Media and world politics. Activism becomes secondary to the media spectacle. The celebrity transcends the grassroots network, displacing the logic of organized networks as new institutional forms. The small cohort of insiders involved with network governance (system administrators, programmers, lawyers, editors, advisory board members) is collapsed into the individual who signs the book deal for a million bucks. For WikiLeaks the trans-institutional relations are not so much between a network of networks, but between a network and broadcast media institutions. The potential network of chapters distributed around the world, based on investigative teams or some other form of collaborative analysis, building on expertise from Wikipedia to Indymedia, did not materialize. Instead, the work of analysis was outsourced to mainstream newspapers.

Whereas civil society organizations such as NGOs tend to play by the rules and seek legitimacy from dominant institutions, WikiLeaks' strategy is a populist one that taps into widespread public disaffection with mainstream politics. Political legitimacy, for WikiLeaks, is no longer something graciously bestowed upon minor actors by the powers that be. WikiLeaks bypasses this old world structure of power and instead goes to the source of political legitimacy in today's info-society: the rapturous banality of the spectacle. The missionary zeal to enlighten the idiotic masses and "expose" the lies of government, the military, and corporations is reminiscent of a media-culture paradigm of the 1940s-50s. Think Adorno, Horkheimer, Lazarsfeld, and later Katz.

The work of interpreting leaked files, very oddly, is left up to the few remaining on-staff journalists in select "quality" news media. Later on, academics pick up the scraps and spin the stories behind the closed gates of publishing stables. But where is the critical networked commentariat? For sure, we are all busy with our minor critiques, but it remains the case that WikiLeaks generates its capacity to inspire irritation at the big end of town precisely because of the transversal and symbiotic relation it holds with establishment media institutions. There's a lesson here for the multitudes – get out of the ghetto and connect with the Oedipal other. Therein lies the conflictual terrain of the political.

Leaving the skepticism about the need to visualize networks aside we should address the politics of code. What is the code that makes possible connections in and across networks? The so-called open culture of networks – derived from the transparent, readily available source code of programmers – is closed in a cognitive sense. The goal of openness is the culture of the club. For the network it is social ties that matter. Networks can be open and grow in all directions. They can also go through an inward-looking phase and strengthen ties. This is what we call organized networks – a process of scalar transformation that institutes social-technical capacities in ways that rival or indeed take over traditional institutions that have defined modern life (government, unions, universities, firms). They remain virtual in that they use the benefits of translocal, global communication, while overcoming the down-side of the famous "weak ties" that are the cause of hyper-growth, but also non-commitment. What we need is destiny design. Software with consequences.

Amidst all the accumulating crises it is clear that we need to do something. The old models of commitment (party, church, movement) no longer appeal to most of us. It is not enough simply to inform, to network. We organize to attack. Networks are created to take initiative, to lead us into new situations, not merely to "keep updated." Twitter shouldn't ask "What's going on?," but "Do you join me?" In this respect, networks are the ground of invention that accommodates internal hacks or a collective capacity to make decisions within techno-cultures. Yet, as Wayne Price notes, "Organizations are obstacles to organizing ourselves."[1] So how would crowdsourcing interpretation be organized in the event of something like WikiLeaks? The aesthetics of representation offers the image of organization, but not a strategy or method of movement and transformation. Returning to the central planning committee special to the Leninist party form, as Jodi Dean and others would have it, does not particularly help either: "*The party must prepare the revolution.* Here the party is producer and product (feedback, networks, self-organization, emergence). It is an exclusive organization that interacts with, and learns from, the struggles and suffering of the people."[2]

There's a form of deep romanticism at work here, grown out of a real-existing despair speaking from inside the Empire in decline, and an analytical failure to think through the logic of the party, the contemporary media environment, and the work of politics as we know it. There's not a question of a lack of critical intelligence or political passions here, but an insufficient understanding of the relation

1   Wayne Price, "Insurrectional Anarchism vs. Class-Struggle Anarchism," 2010, HTTP://WWW.ANARCHISTNEWS.ORG/?Q=NODE/12729.

2   Jodi Dean, "Primer on the Leninist party," October 11, 2010, HTTP://JDEANICITE.TYPEPAD.COM/I_CITE/2010/10/PRIMER-ON-THE-LENINIST-PARTY.HTML.

between networked media and forms of self-organization and politics. Again, we see a return to the logic of representation, organizational form, and politics. Despite softening the claim to engage "the struggles and suffering of the people" with the appeal to "feedback, networks, self-organization, emergence," at its heart the structure of the party can only ever be about representation. And at this point politics vacates the territory of organizational form, since the logic of networks is about relations not representations, processes not procedures.

We find more purchase in the work of architect Keller Easterling with her interest in hidden organizations, global infrastructures, the production of protocols, and the role of multipliers: "Perhaps because these organizations operate in the background, in an active and relational rather than nominative register, their political outcomes are often at once pervasive and mysterious."[3] What we need to understand is how today's networks are (dis)organizing us. What is the mystery of these invisible, still unknown protocols that shape our social life that we fail to grasp? And how do we register the distribution of passions and their mobilization of politics across seemingly immaterial networks? These are the key tasks for the work of organization today.

The Network will not be Revolutionized

Welcome to the politics of diversion. There is a growing paradox between the real existing looseness, the "tyranny of structurelessness" on the one hand, and desire to organize in familiar structures such as the trade union, party, and movement on the other. Both options are problematic. Activists, especially those from the baby-boomer generation, do not like to speculate on the potential of networks as they fluctuate too much – an anxiety perhaps fueled by the instability of their pension funds. Networks are known for their unreliability and unsustainability. Even though they can scale up in unprecedented ways, and have the potential to perform real-time global politics from below, they also disintegrate in the same speed. Like Protestant churches and Christian sects, leftist political parties and traditional union structures can give people a much needed structure to their life. It is hard to argue against the healing, therapeutic value that such

3   Keller Easterling, *Extrastatecraft: Hidden Organizations, Spatial Contagions and Activism — Research Project*, HTTP://WWW.JANVANEYCK.NL/0_3_3_RESEARCH_ INFO/DESIGN_EXTRASTATECRAFT.HTML.

organizations can have on societies and neighborhoods that are under severe pressure of disintegration. What we observe is that these two strategies are diverging models. They do not compete, but they do not necessarily overlap either.

Let's dream up an Indymedia 2.0. No more Wikipedia neutrality or WikiLeaks male celebrity blues. Where are the social networking sites for activists? The internet flagship of the "other globalization movement," Indymedia, has not changed since its inception in late 1999. Of course its website content has grown – there are now editions in dozens of languages, with a variety of local and national nodes that we rarely see on the Net. But the conceptual basics are still the same. The move to smart phones wasn't made. The problems have been identified a long time ago: there is an ongoing confusion between the model of the alternative news agency, the practical community organization level, and strategic debates. All too often Indymedia is used as an "alternative CNN." There is nothing wrong with that, except that the nature of the corporate news industry itself has changed fundamentally. Why was the crucial migration to an independent social media platform not made? This is now a question that future social movement historians will have to deal with.

What does Indymedia 2.0 mean? The question of why indymedia. org failed and did not further develop into an active and open social networking site or clearly take up a position in the Web 2.0 debate is something that needs to be addressed.[4] Have media activists already learnt enough of the Brechtian Indymedia *Lehrstück* that started in the late nineties? Is global branding and branching, as in the case of Indymedia (one name, often similar design, sharing of servers, some syndication of content, etc.), still as important as it used to be? The original Indymedia concept consisted of a mix of forum software with asynchronous threads and early blog features as later developed by WordPress. Indymedia met the challenge of scalability in amazing ways only to discover its limits. Although Indymedia replicated itself into regional sites in various languages, it did not make the move to turn itself into collaborative blog software. The dominant Web 2.0 blogging ideology emphasized the empowerment of the individual

---

4   See the nettime mailing list debate, "If Only Indymedia Learnt to Innovate," November 2008, HTTPS://NETTIME.ORG/LISTS-ARCHIVES/NETTIME-L-0811/ THREADS.HTML.

user. Indymedia, on the other hand, was a collective, collaborative site that involved users in the news gathering. Blog software, on the other hand, worked on the premise of distributed links ("blogrolls"). The blog community did indeed exist, however, it was dispersed by design. Contamination seems key for transnational social-political networks. As do regular face-to-face meetings. Let your network connect with the concrete and adaptation and transformation will undoubtedly kick in. Then try reconnecting across networks (and other institutional and organizational forms) on the global scale. Conflict will already have multiplied and the primary condition of sustainability will be underway.

Governance requires protocols of dissent. The governance of networks is most clearly brought into question at the borders of networks. Control is the issue here. Borders function to at once regulate entry, but they also invite secret societies to infiltrate by other means. The contest between these two dynamics can be understood as the battle between governmental regimes and non-governmental desires. We do not have to decide here as we have split agendas: we long for order in times of chaos and simultaneously overload and dream of free information streams. This brings us to the related issue of sustainability. If the borders of networks consist of governmental and non-governmental elements (administration versus inspired sabotage and the will to infiltrate), then we can also say that the borders of networks highlight their inherent fragility. How can this be turned into a strength for the future of networks? There are always overlaps of identity and social structures.

The revolution will be participatory or she will not be. If there is no desire addressed, not much will happen. YouTube and Facebook are fueled with no shortage of desire. Rightly or not, they are considered the apogee of participatory media. But they are hardly hotbeds of media activism. Linux geeks – leave the ecosphere of servicing free software cartels. The abbreviation policy, from G8 to WTO, has failed, precisely because abstract complex arrangements within global capitalism do not translate well into the messy everyday. By contrast, the NGO movements, at their best (we won't go into a catalogue of failures here), have proven the efficacy of situated networks. The problem of trans-scalar movement, however, remains. This was made clear in the multi-stakeholder governance model adopted by government, business and civil society organizations throughout the UN's World

Summit on the Information Society (2003–2005). Here we saw a few civil society organizations find a seat at the negotiating table, but it didn't amount to much more than a temporary gestural economy. As civil society participants scaled the ladder of political/discursive legitimacy, the logic of their networks began to fade away. This is the problematic we speak of between seemingly structureless networks and structured organizations. The obsession with democracy provides another register of this social-technical condition.

The borders of networks are the spatial sites of politics. As networks undergo the transversal process of scalar transformation, the borders of networks are revealed as both limits and possibilities. In the process of growth the kernel of a network crystallizes a high energy. After some months or, for the lucky ones, a few years, there is no longer an inside of networks, only the ruins of the border. This is an enormous challenge for networks – how to engage the border as the condition of transformation and renewal?

The borders of networks comprise the "'non-democratic' element of democracy" (Balibar/Mezzadra). This insight is particularly helpful when thinking the political of networks, since it signals the fact that networks are not by default open, horizontal, and global. This is the mistake of much of the discourse on networks. There is no politics of networks if there are no borders of networks. Instead of forcing democracy onto networks, either through policing or installed software, we should investigate its nature. This does not mean that we have to openly support "benevolent dictatorships" or enlightened totalitarian rule. Usually networks thrive on small-scale informality, particularly in the early existence of their social-technical structures.

There are no citizens of the media. Find and replace the citizen with users. Users have rights too. The user is not a non-historical category but rather a system-specific actor that holds no relationship to modernity's institutions and their corresponding discourse on rights. What is needed, then, is total reengineering of user-rights within the logic of networks. As much as citizen journalists, liberal democratic governments, big media, and global institutions are endlessly effusive about their democratic credentials, organized networks are equally insistent in maintaining a "non-democratic" politics. A politics without representation – since how do networks represent anything? – and instead a non-representational politics of relations. Non-democratic does not mean anti-democratic or elitist. It has proven of strategic importance to loosen ties between "democracy" and "the media." Let's remember that the citizen journalist is always tied to the media organs of the nation-state. Networks are not nations. In times of an abundance of channels, platforms, and networks, it is no longer necessary to claim "access." The democratization of the media has come to an end. People are tired of reading the same old critique of the *New York Times*, CNN, and other news outlets that are so obviously Western and neoliberal biased. It is time to concentrate our efforts on the politics of filtering. What information do we want to read and pass on? What happens when you find out that I am filtering you out? Do we only link to "friends"? And what to make of this obsessive compulsion to collect friends? Would it be alright if we replaced friends with comrades? What could object against the tendency to build social networks? Wasn't this what so many activists dreamt of?

Scalar Relations

Applied scalability is the new technics. How to crack the mystery of scalability and transformation of issues into a critical proliferation of protest with revolutionary potential? With the tendency of networks to regress into ghettoes of self-affirmation (the multitudes are all men), we can say that in many ways networks have yet to engage the political. The coalition building that attends the process of trans-scalar movement will by design create an immanent relation between networks and the political. Moreover, it will greatly facilitate the theoretical and analytical understanding of networks. Tension precipitates the will to utterance, to express, and to act. And it is time for networks to go to work.

Let's turn, now, to perhaps least investigated aspect of scalability. Why is it so difficult for networks to scale up? There seems to be a tendency to split up in a thousand micro conversations. This also counts for the "social software" blogs like Orkut, Friendster, and LinkedIn, in which millions from all over the globe still participate. For the time being it is only the geeky Slashdot that manages to centralize conversations amongst the tens of thousands of its online users. Electronic mailing lists do not seem to get above a few thousand before the conversation actually slows down, heavily moderated as it is. The ideal size for an in-depth, open discussion still seems to be somewhere between 50 and 500 participants. What does this mean for the networked multitudes? To what extent is this all a software issue?[5] Could the necessary protocols be written up by women? Well, of course, but what protocols would be adopted in such a case? Can we imagine very large-scale conversations that do not only make sense but also have an impact? What network cultures can become large transformative institutions?

Perhaps organized networks will always remain virtual. This option should never be dropped. There is no secret plan to institutionalize in the brick and mortar world. Maybe organized networks cannot work in collaboration with existing institutional structures. If so, how might the virtual be formalized? By this we don't mean formalization in the old sense whereby the network takes on a hierarchical structure made up of a director, an elected secretariat, and so forth. Such a model was adopted by the grassroots movements of the 1960s and 70s, and is now the primary reason why such entities are unable to deal with the demands and realities of networked sociality. Against this mode of formalization, how might informality acquire an organized response to the unpredictability of needs and crisis and the rhythms of global capital?

As unstable as this model may sound, perhaps it is the form best suited to the habitus of networks. It is necessary, after all, to identify the characteristics, tendencies, and limits – that's to say, the short history – of the network, and develop a plan from there. There's no point assuming that established patterns of communication and practice can somehow be evaporated and entirely new projects started afresh. To do so would mean the invention of a new network, and that would mean undertaking that time-consuming work of defining practices and protocols through experimentation, trial and error. By all means, let's see new networks emerge – they will in any case. But the solution is not to abandon the hard labor, accumulated resources, and curious network personas – or brand, if you like – that have already been cultivated. Let's take the next step.

While it seems that we're forever in some perpetual crisis and phase of transition, now really is the time for the organized network to establish the ground upon which new politics, new economies, and new cultures may emerge within the dynamics of the social-technical

---

5   Here we're thinking of collaborative, peer-to-peer "software solutions" such as Paper Airplane, HTTP://PAPERAIRPLANE.US. Thanks to Soenke Zehle for bringing this site to our attention.

system. In this way, the network opens up to an entirely new range of external variables that in turn function to transform the internal operation of the network. Such is the work of the constitutive outside – a process of post-negativity in which rupture and antagonism affirms the future life of the network. The tension between internal dynamics and external forces comprise a new ground of the political.

Radical democracy theorists are still so slow and far away from recognizing this new field of techno-sociality. Where they posit a negation of social antagonisms within ideologies such as the Third Way, and thus identify the disintegration of liberal democratic principles, the emergence of organized networks, by contrast, are constituted precisely in this denial of antagonisms by the culture of liberal democracy. The institutional structures of liberal democracy have become disconnected from the field of sociality, and in so doing are unable to address the antagonisms of the political. Antagonisms do not evacuate the scene so much as take flight into new terrains of communication. The organized network is open to the antagonisms that comprise social-technical relations. For this reason, it is urgent that organized networks confront the demands of scale and sustainability in order to create new institutional horizons within which conflicts find a space of expression and a capacity for invention.

Accompanying such a transformation is the recognition of power structures and the fact that organized networks will always be shut out of them. There are also internal informal power structures – a recognition of which is the first step towards transparency. Too often the denial of existing structures prevents a discussion of how new forms of organization could emerge. The prevailing assumption of decentralization shuts down debate and imagination of how things could be done differently. Moreover, it reproduces the absolute power of the geeks. For them, it's not an issue because they can safely continue their engineering class without having to confront the urgency of translation that accompanies networks seeking to deal with the turmoil of new socialities.

Similarly, the structures that call themselves networks deny how centralized they are. Here, we are thinking of the proliferation of "research networks" within universities. There is an amazing confusion about what networks are within these settings. In many ways, such obfuscation is quite deliberate: since the institution of the university – a networked organization – is beyond repair and unable to deal with the complexities of an informatized society, it is no wonder that we see this latest attempt at window dressing. There is a bizarre assumption that if governments and funding bodies throw money at projects that demonstrate a correspondence with networks – whatever that might mean – then, by some peculiar magical process, "innovation" (another quite meaningless term) will emerge. And what do you know, the procedure for submitting proposals, developing research partners, justifying budgets, outlining time schedules, undertaking research, and so on and so forth is exactly the same as the previous year of harvesting. The result: the existing elites are rewarded, and power is consolidated through the much more accurate model of the "cluster" (a rather ugly word that finds its birthplace in the school playground). There is no chance for these so-called networks to encounter infection. Quarantined inquiry is what these research networks are all about. Why? Because there is a complete failure to engage the technics of communications media in the first instance, to say nothing of the dependency model of funding which simply functions to reproduce the same.

Social Media Critique

The internet turns out to be neither the problem nor the solution for the global recession. As an indifferent bystander it doesn't lend itself easily as a revolutionary tool. The virtual has become the everyday. The New Deal is presented as green, not digital. The digital is a given. This low-key position presents an opportunity to rethink the social media hype. How might we understand our political, emotional, and social involvement in internet culture over the next few years?

News media is awash with "economic crisis," indulging in its self-generated spectacle of imminent financial meltdown. Experts are mobilized, but only to produce the drama of dissensus. Programmed disagreement is the consensus of daily news. Crisis, after all, is the condition of possibility for capitalism. Unlike the dotcom crash in 2000–2001, when the collapse of high-tech stocks fueled the global recession, the internet has so far managed to stay out of the blame game. Social media networks only suffer mild side effects from the odd collection of platforms and services, from Google to Wikipedia, Photobucket, Craigslist, MySpace, Facebook, Twitter, Habbo, and so-called regional players such as Baidu, Weibo, Alibaba, Toudou Youku, Tencent, and 51.com. Despite its benign existence, there still is hyper-growth wherever you look. Apps and platforms remain "new" but show a tendency to get lost inside the boring, stressful, and uncertain working life of the connected billions.

Social media networks are technologies of entertainment and diffusion. The social reality they create is real, but as a technology of immediacy you can't get no satisfaction. We initially love them for their distraction from the torture of now-time. Networking sites are social drugs for those in need of the Human that is located elsewhere in time or space. It is the pseudo Other that we are connecting to. Not the radical Other or some real Other. We systematically explore weakness and vagueness and are pressed to further enhance the exhibition of the Self. "I might know you (but I don't). Do you mind knowing me?" The pleasure principle of entertainment thus diffuses social antagonisms – how does conflict manifest within the comfort zones of social networks and their tapestries of auto-customization? The business-minded "trust doctrine" has all but eliminated the open, dirty internet forums. Most social media are echo chambers of the same old opinions and cultural patterns. As we can all witness, they are not exactly hotbeds of alternative sub-culture. What's new are their "social" qualities: the network is the message. What's created here is a sense or approximation of the social. Social networks register a "refusal of work." But our net-time, after all, is another kind of labor. Herein lies the perversity of social networks: however radical they may be, they will always be data mined. They are designed to be exploited. Refusal of work becomes just another form of making a buck that you never see.

Once social networking sites were fashion victims as much as everything else. They used to come and go. Before 2008 the dominant dynamic could have been summarized like this: The moving herds that go from one server to the next demonstrate an impulsive grazing mentality: once the latest widgets are installed, it's time to move on. The speed of migration across space signaled the de-culturization of software. While Orkut disappeared in G8 countries, it used to be big in Brazil. After 2008 these regional differences started to disappear. Once the global masses were locked-in, we can only entertain nostalgic dreams of a return of the fashion principle in social media. Is anyone still seriously investing in real estate in Second Life? Of course not. What the online world needs is sustainable social relations outside of the monopoly platforms.

Let's look at the lessons from the major social movements over the last 50 years. The force of accumulated social-political desires manifest, eventually, in national and global forums that permeate back into policy discourse and social practice: think March on Washington, 1963 (Black Civil Rights), Rio, 1992 (Earth Summit), Porto Alegre, 2001 (World Social Forum), Geneva and Tunis, 2003–2005 (World Summit on the Info-Society). None of these examples are exempt from critique. We note them here to signal the relationship between sustainability and scalar transformation. We are familiar with formats such as bar camps, un-conferencing, and have participated in DIY techno-workshops at those seasonal media arts festivals. But these are hardly instances of sustainability. Their temporality of tinkering is governed by the duration of the event. True, there is occasionally resonance back in the local hack-lab, but such practices are exclusive to techno-secret societies, not the networked masses. Social networking sites are remarkable for their capacity to scale. Their weakness is their seeming incapacity to effect political change in any substantive way. The valorization of citizen-journalism is not the same as radical intervention, and is better understood as symptomatic of the structural logic of outsourcing media production and election campaign management.

"From social to socialism is a small step for humankind, but a big step for the Western subject." (John Sjerpstra)

What makes the social attractive, and socialism so old school and boring? What is the social anyway? We have to be aware that such post-modern academic language games do not deepen our understanding of the issues, nor widen our political fantasies. We need imagination, but only if it illuminates concepts that transform concrete conditions. The resurrection of the social after its disappearance is not an appealing slogan. Some ideas have an almost direct access to our body. Others remain dead. This in particular counts for insider jargon such as rent, multitude, common, commons, and communism. There's a compulsion to self-referentiality here that's not so different from the narcissistic default of so many blogs. What, then, are the collective concepts of the social networked masses? For now, they are engineered from the top-down by the corporate programmers, or they are outsourced to the world of widgets. Tag, Connect, Friend, Link, Share, Tweet. These are not terms that signal any form of collective intelligence, creativity or networked socialism. They are directives from the Central Software

Committee. "Participation" in "social networks" will no longer work, if it ever did, as the magic recipe to transform tired and boring individuals into cool members of the mythological Collective Intelligence. If you're not an interesting individual, your participation is not really interesting. Data clouds, after all, are clouds: they fade away. Better social networks are organized networks involving better individuals – it's your responsibility, it's your time. What is needed is an invention of social network software where everybody is a concept designer. Let's kill the click and unleash a thousand million tiny tinkerers!

We are addicted to ghettoes, and in so doing refuse the antagonism of the political. Where is the enemy? Not on Facebook, where you can only have "friends." What social media lacks is the technique of antagonistic linkage. Instead, we are confronted with the Tyranny of Positive Energy. Life only consists of uplifting experiences. Depression is not a design principle. Wikipedia's reliance on "good faith" and its policing of protocols quite frequently make for a depressing experience in the face of an absence of singular style. There is no "neutral point of view." This software design principle merely reproduces the One Belief System. Formats need to be transformed if they are going to accommodate the plurality of expression of networked life. Templates function as zones of exclusion. But strangely, they also exclude the conflict of the border. The virus is the closest thing to conflict online. But viruses work in invisible ways and function as a generator of service labor for the computer nerd who comes in and cleans your computer.

The critique of simulation falls short here. There is nothing "false" about the virtuality of social networking sites. They are about as real it gets these days. Stability accumulates for those hooked to networks. Things just keep expanding. More requests. More friends. More time for social-time. With the closure of factories comes the opening of data mines. Privacy is so empty of curiosity that we are compelled to slap it on our Wall for all to see. If we are lucky, a Friend refurbishes it with a comment. And if you are feeling cheeky, then Throw A Sheep! You would be hard-pressed to notice any substantive change. But you will be required to do never-ending maintenance work to manage all your data feeds and updates. That'll subtract a bit of time from your daily routine.

Social media is not for free. "Free as in free beer" is not like "free as in freedom." Open does not equal free. These days "free" is just another word for service economies. The Linux fiefdom know that all too well. We need to question naïve campaigns that merely promote "free culture" without questioning the underlying parasitic economy and the "deprofessionalization" of cultural work. Pervasive profiling is the cost of this opening to "free market values." As users and pro-sumers we are limited by our capacity as data producers. Our tastes and preferences, our opinions and movements are the market price to pay. At present, Facebook's voluntary and enthusiastic auto-filing system on a mass scale represents the high point of this strategy. But we cannot succumb to the control paranoia and to the logic of fear. Let's inject more kaos in it! So what if you have your anti-whatever Facebook group? What does it change other than expanding your number of friends? Is deleting the radical gesture of 2018? Why not come up with a more subversive and funny, anti-cyclical act? Are you also looking for rebel tactical tools?

Soon the social media business model will be obsolete. It is based on the endless growth principle, pushed by the endless growth of consumerism. The business model still echoes the silly 90s dotcom model: if growth stagnates, it means the venture has failed, and needs to be closed down. Seamless growth of customized advertising is the fuel of this form of capitalism, decentralized by the user-prosumer. Mental environment pollution is parallel to natural environment pollution. But our world is finished (limited). We have to start designing appropriate technologies for a finite world. There is no exteriority, no other worlds (second, third, fourth worlds) where we can dump the collateral effects of insane development. We know that Progress is a bloodthirsty god that extracts a heavy human sacrifice. A good end cannot justify a bad means. On the contrary, technologies are means that have to justify the end of collective freedom. No sacrifice will be tolerated: martyrs are not welcome. Neither are heroes.

"Better a complex identity than an identity complex" (Jo van der Spek). We need to promote peer-education that shifts the default culture of auto-formation to the nihilist pleasure of hacking the system. Personal exhibition on social media networks resembles the discovery of sexuality. Anxiety over masturbation meets digital narcissism (obsessive touching up of personal profiles) and digital voyeurism (compulsive viewing of other's profiles, their list of friends, secrets, etc.). To avoid the double trap of blind techno-philia and luddite techno-phobia, we have to develop complex digital identities. They have to answer to individual desires and satisfy multiple needs. Open-ID are a good starting point. "Steal my profile." It's time to remix identity. Anonymity is a good alternative to the pressures of the control society, but there must be other alternatives on offer. One strategy could be to make the one ("real") identity more complex and, where possible, contradictory. But whatever your identify might be, it will always be harvested. If you must participate in the accumulation economy for those in control of the data mines, then the least you can do is Fake Your Persona.

9.

THE ART OF COLLECTIVE COORDINATION

We urge everyone in our district to organize for justice –
then cars will not be burnt, stones will not be thrown.
– Community activist groups Megafonen and Pantrarna,
Aftonbladet, Stockholm, May 2013.

Why do folks keep expecting technology to fix social
issues that society hasn't been able to fix?
– danah boyd, Twitter, May 29, 2013.

Why not? is a powerful question and something you
should ask every day.
– Eli Broad, *The Art of Being Unreasonable*, 2012.

INTERNET ACTIVISM HAS GROWN UP AND IS HUGE, COMPARABLE TO gender struggles and climate change disputes. This is the age of WikiLeaks, Anonymous, denial-of-service attacks on vital infrastructure, and National Security Agency whistle blower Edward Snowden, all capturing the global imagination – a world which, already for decades, has remained *terra incognita* for the (media) establishment. The right to communicate is vital and no longer a luxury. Yet the revolutionary spread of connectivity and storage does not translate into an equivalent victory for the freedom of communication. Quite the contrary. After a good decade of struggles since 9/11, cyber-rights activists are in danger of falling into a lethargic state of depression. A picture emerges of a globe with increasing connectivity and a growing diversity of crises, with short-lived protest movements that accompany a loss of legitimacy of the political classes. The question posed here is whether small and dense communities can be a possible answer to the crisis of the family, church, trade unions, and political parties as traditional social formations. If sit-ins, affinity groups, alternative scenes, and autonomous cells are phenomena of the past, can organized networks become the preferred forms of sustained political mobilization for the decade to come?

We are all still struggling to make sense of what happened during 2011, the "belated" Year of Protest that started off with the Arab Spring and culminated in the Occupy movement, so neatly summarized by Slavoj Žižek in his book *The Year of Dreaming Dangerously*. We can ask ourselves why it took three to four years for these events to unfold – and why we are, in retrospect, already six years underway interpreting these chains of global events. Why didn't 2011 culminate in a larger political momentum? Did we need a break (a "Pause for the People"?) during 2012, before a next wave of protests could begin again in Bulgaria, Sweden, Turkey, Brazil, and Egypt?[1] As David DeGraw concluded: "Through Anonymous, Occupy and the 99% Movement, we collectively proved that decentralized self-organizing networks of like-minded people rallying together can set the world on

---

1   Writing about the "unnatural relative calm of the spring of 2012," Slavoj Žižek observes, "[W]hat makes the situation so ominous is the all-pervasive sense of blockage: there is no clear way out and the ruling elite is clearly losing its ability to rule." *The Year of Dreaming Dangerously* (London: Verso Books, London, 2012), 197.

fire. However, we lacked an exit strategy and the resources required to build a self-sustaining movement that can truly achieve the change and evolution of society that we all know we need."[2] This discussion is by no means limited to the (overstated) role of social media and mobile phones in these mass mobilizations. We have to ask ourselves: what does this hermeneutic delay mean in an age of real-time digital networks where events (including the instant interpretations of Žižek and other public intellectuals) travel at the speed of light?

Extensive use of smart phones seems to make it even harder for activists to reflect on the range and impacts of their actions. The "walled gardens" of Facebook and Twitter (also called "echo chambers") make it hard to estimate the real scale and impact of one's semi-private conversations (including the public "clicktivism" à la Avaaz). Is "direct action" becoming even more symbolic (and informational) than it already was? The integration of urban space and digitally networked space is a fact, and the Movement of the Squares (from Tahrir to Taksim) is first to admit this techno-condition. Besides a few internet gurus such as Clay Shirky and Jeff Jarvis, who can read such "Facebook revolutions" only as giant input devices as expressions of "citizen journalism," executed in the name of US-American (market) values, there is little patience among the *feuilleton* writers to study in detail what is going on here (exceptions being Zeynep Tufekci and Eric Kluitenberg, among others). Can we speak of a theoretical deficit or rather an overproduction in terms of reporting? Social grooming and the "presentation of the self" may be sociological facts in research on the topic, but they say surprisingly little about organizational questions of sociality.[3] Be it stylish, aggressive, desperate, or diplomatic, the self-promotion on dominant social media platforms such as Facebook and Twitter has become, in essence, part of the broadcasting logic of old media: messaging the void.

The concept and practice of organized networks outlined in this book is merely a proposal, a possible answer to how we might

2　David DeGraw, "The Manhattan Project for the Evolution
　of Society," May 20, 2013, HTTP://DAVIDDEGRAW.ORG/
　MANHATTAN-PROJECT-FOR-THE-EVOLUTION-OF-SOCIETY/.

3　See the range of literature, varying from the 1959 classic of Erving Goffman,
　The Presentation of the Self in Everyday Life (New York: Anchor Books, 1959),
　to contemporary internet scholars such as Nancy Baym and danah boyd.

overcome the insular status of the subject-as-user at a moment when traditional institutions such as political parties, unions, and Western parliamentary constitutions are in crisis. Along with other long-standing institutional forms such as firms and universities, unions and political parties have not, of course, gone untouched by the impact of digital communication systems on how organizations manage, devise, and coordinate their internal operations and external campaigns. Yet they do so as *networked organizations*, which are notable for the ways in which their practices remain largely bound to the cultural logic of hierarchical organization. Organized networks, by contrast, tend to be more horizontal and emerge from within the technics of digital communication media. This does not mean they are free from issues of power and hierarchy (no socio-technical organizational form is). But it does give orgnets a degree of flexibility, spontaneity, and scalability that is beyond the capacity of most networked organizations. This "born digital" feature of organized networks is also the cause of a range of constraints, even weaknesses: longevity and continuity over time are frequently a problem; financing operations, projects, and practices beyond a shoe-string or zero budget is always a challenge; similarly, orgnets have a general dependency on free and voluntary labor, which is another reason their longevity is typically uncertain and insecure; and not adopting technically stable software that is well designed and maintained over time is something that bedevils pretty much all software alternatives for social media.

Needless to say, organized networks are proliferating across the planet, manifesting in a range of projects and practices championed by non-profit organizations, activist networks, cultural organizations, alternative schools, and research groups. All seek to build foundations to support their interests and agendas outside the architectonic form of brick and mortar. This doesn't mean orgnets are purely "virtual," online formations. Flesh meets are crucial, and this often requires drawing on the institutional resources of networked organizations. And, perhaps most importantly, orgnets are galvanized by political passions that are leftist in orientation. This is what separates them from similarly organized social-technical forms: tech start-ups, think tanks, criminal syndicates, and terrorist networks. But can we identify the catalyzing force or event that compels bodies and brains to become organized?

There is no longer the need for a call to get involved. Discontent is thriving. We can be glad that the Age of Indifference is over. But how do we shape today's solidarity? Is it only a matter of "capturing" and "channeling" political energies that are floating around us? No matter where one looks, one gets the feeling that urgency is only in its infancy and that (social) networks have not even remotely started to explore their full potential as organizational machines, discursive platforms, and desire tools. In the spirit of the Unlike Us initiative we can only say: Unfriend Facebook and restart your network imaginary.[4] Without falling into the romantic trap of some harmonious offline life, the Unlike Us research network asks what sort of network architectures could be designed that contribute to "the common," understood as a shared resource and system of collective production that supports new forms of social organizations (such as organized networks) without mining for data to sell. Or, as network theorist Tiziana Terranova puts it, "the concept of the common is used here as a way to instigate the thought and practice of a possible post-capitalist mode of existence for networked digital media."[5]

Activism never restricted itself to the slow and invisible process of advocacy. It is boring to become a citizen again, stand up, and criticize mainstream media. We left the 20th century ages ago. "Movimentalism" is not a preliminary form of "collective awareness" either. In many forms of institutional politics the role of "civil society" is reduced to that of an input device: "Thanks, we got your message, now shut up." This contradicts another neoliberal adage, which says that citizens should not complain but instead "embody" solutions (and not merely suggest them). We only have a right to complain if we have alternatives on hand that demonstrably work. Current political bureaucracies can no longer deal with anger. This processual numbness and lack of patience in turn infuriates the popular voice. Another result is repression for nothing, outbursts of excessive show of force, and out-of-the-blue violence by authorities that no one seems to be able to explain. To contradict authority in public these days all too easily leads to arrests or, at worst, to the killing of protesters by police.

---

4   Inaugurated in 2011, the Unlike Us network of artists, designers, scholars, activists, and programmers undertakes research on "alternatives in social media." See HTTP://NETWORKCULTURES.ORG/UNLIKEUS/.

5   Tiziana Terranova, "Red Stack Attack! Algorithms, Capital, and the Automation of the Common," in Robin Mackay and Armen Avanessian (eds), *#Accelerate: The Accelerationist Reader* (Falmouth: Urbanomic, 2014), 382.

Resistance grows out of an existential crisis. This also counts for net.activism, which expresses itself first and foremost as informal resentment in semi-closed private conversations. Whether enclosed within the gated communities of social media or "hidden" within the obscure world of activist mailing lists, we now know, following the Snowden revelations, that such exchanges are available for inspection. There is violence, lack of housing, unemployment, pollution. And all this is recorded, shared, and stored online. Amid this atmosphere of growing agitation, taking action is no longer a gesture of boredom or prosperity. Activists have fires to combat. Yet urgency in itself does not easily translate into a specific political form. We need to re-invent these political forms, time and again – and this is where the role of designers, artists, educators, curators, and critics comes in. But where are they, now that the network needs them? Hiding in their ateliers, classrooms, offices, and galleries, waiting for the digital storm to pass by? How are we to understand the idea of the contemporary when it may take years, decades even, for the distribution of concepts to find traction as infrastructural forms on a mass scale? What role in all of this do think tanks, hackers, and funding bodies play? Have you already seen mobile WiFi research units supported by portable offline libraries stored on USB devices? Not sexy for your museum, perhaps? It will happen, regardless. As Alex Williams and Nick Srnicek argue in their "Accelerate Manifesto," "we need to build an intellectual infra-structure."[6] And this is why we believe artists, educators, and their ilk have a key role in designing alternative infrastructures of communica-tion, knowledge production and social-political organization. Why? Because the technical is not independent of the conceptual, and for far too long it has been assumed that infrastructure design can be left up to the engineers and computer programmers. As we noted at the start of this book, this is a serious strategic mistake. We have seen for some time now the ways in which concepts first developed among ac-tivist networks migrate as memes into policy discourse associated with the creative industries and urban planning. To design robust concepts for infrastructure, no matter how imaginary they may be, is to stake a

6   See Alex Williams and Nick Srnicek, "#Accelerate Manifesto
     for an Accelerationist Politics," posted on May 13, 2013,
     HTTP://CRITICALLEGALTHINKING.COM/2013/05/14/
     ACCELERATE-MANIFESTO-FOR-AN-ACCELERATIONIST-POLITICS/.

claim on the production of the future-present. Science fiction writers have been doing this for decades. We note McKenzie Wark's injunction "to build prototypes for another life in the margins."[7] It is only a matter of time before authorities in need of an idea do the work of scaling them up. The indexicality of the concept will remain inscribed within the technical parameters of organizational infrastructures. Take this as a form of technological determinism, if you will.

Think sustainable networks that spread progressive knowledge, with strong ties between all countries and continents. Yes, there is the obligation to represent and build larger structures, but the avalanche of catastrophic occurrences only seems to grow. Initially an impulse, activism nowadays quickly mutates into a daily informational routine. The problem is one of neither consciousness nor commitment but of the organizational form through which we express our discontent. This explains the shift in attention not only toward political parties, such as the Italian Five Star Movement and pirate parties in a range of European countries (notably Sweden and Germany) but also toward concepts like organized networks, as well as the Multitude (Hardt and Negri), the critique of horizontalism and communism 2.0 of Jodi Dean, and the emergence of net.political entities such as Anonymous, Avaaz, WikiLeaks, and their shadow internet spectacle, Kony 2012.

Most critiques of the current impasses are known, justified, and predictable. Yes, movements such as Occupy "expend considerable energy on internal direct-democratic process and affective self-valorization over strategic efficacy, and frequently propound a variant of neo-primitivist localism, as if to if to oppose the abstract violence of globalized capital with the flimsy and ephemeral 'authenticity' of communal immediacy," as Williams and Srnicek argue. But while this critique may be valid for US activism, it doesn't seem to resonate with the situation in Southern Europe and the Middle East. Activism in the North-west of Europe, in fact, needs more discussion, more consensus, in order to strengthen its own social if not communal ties, which tend to shut down the possibility of subversion and dissent through group pressures to conform. The problem with Occupy was not its obsession with internal decision-making rituals but rather the limited political capacity of its members to build coalitions. The issue

---

7   McKenzie Wark, "Designs for a New World," *E-Flux Journal* 58 (October 2014), HTTP://WWW.E-FLUX.COM/JOURNAL/DESIGNS-FOR-A-NEW-WORLD/.

is also one of a lifestyle trap. When activism promotes itself as a counter-culture, the ability for its memes to travel outside of the issue context remains limited. Networked politics faces a similar problem: how can we get rid of its Californian hipster dotcom image and politicize the masses of unemployed young people across the globe who will never benefit from the mega profits of "their" Google and Facebook? The "social media" issue is too big, and too strategic, to be left to the stagnating "new media" sector. When will we see the first strike against Free & Open services by its users?

Activism is about saying, "Enough is enough, we've got to stand up and do something." The refusal is foundational. Just Say No. For the positivist managerial class this is the hard part as they would rather skip the schizo-dimension of today's society and prefer instead to deal with reasonable and balanced people. It is true that the despair of the rebel often ends up in a catastrophic, violent event, one that will be overdetermined by the agenda of others. So what is aesthetic negativity in the age of smart phones? We cannot run away from this question. Is there a pure form of techno-nihilism that is both creative and destructive? How can the hacker identity be taken out of the libertarian context? The Anonymous identity design is a promising start in this respect.

Organized Networks as Basic Units

Organization, which is, after all, only the practice of
cooperation and solidarity, is a natural and necessary
condition of social life.
 – Errico Malatesta, "Anarchism and Organization,"
 1897.

I'm about convinced now that there is need for a
new organization in our world. The International
Association for the Advancement of Creative
Maladjustment – men and women who will be as
maladjusted as the prophet Amos.
 – Martin Luther King, 1963.

The regained power of negativity alone cannot explain the sudden rise of ultra-large-scale movements that seem to come out of nowhere. In a variation of what Corey Robin writes in *The Reactionary Mind*, we could observe that in these post–Cold War times social movements do not primarily desire and demand but rather express a culture of loss. They mourn a lost future and show their collective desperation about the lack of public infrastructure and facilities (education, health care), the disappearance of secure jobs, and the prospect of a life without income security. In short, they respond to the global demise of middle-class aspirations resulting from rising socio-economic disparities (in comparison to the 1%). Protesters ask us not to obey them but to feel sorry for them. They want to be seen as glorious losers, acting out their loss, celebrating their victim status. Such appeals are enmeshed with the logic of subtraction without transformation. The protesting persona becomes an archetype that, paradoxically, is unable to act. The 1990s presumption of default harmony, with its "unwillingness to embrace the murky world of power and violent conflict, of tragedy and rupture," is long gone.[8]

We can ask ourselves: what comes next, after boredom has left the scene? David Foster Wallace accurately described the previous phase of middle-class harmony in the 1990s. In his essay on the 2000 McCain campaign, he asks why young voters are so uninterested in politics. He observes that it is "next to impossible to get someone to think hard about why he's not interested in something. The boredom itself pre-empts inquiry: the fact of the feeling's enough." This is also true of observations Jean Baudrillard has made concerning "the strength of inertia," the Silence of the Masses.[9] Occasionally wild and spontaneous, though mostly subsisting within a passive neutrality of indifference, the masses are what remain following the evacuation of the social. But this state of affairs one day comes to an end. Once the party has started, it is hard to remain on the sidelines. Wallace notices that politics is not cool. "Cool, interesting, alive people do not seem to be the ones who are drawn to the political process."[10] There is a "deep disengagement that is often a defense against pain.

8   Corey Robin, *The Reactionary Mind* (New York: Oxford University Press, 2013), 172–73.

9   Jean Baudrillard, *In the Shadow of the Silent Majorities . . . Or the End of the Social and Other Essays*, trans. Paul Foss, Paul Patton, and John Johnson (New York: Semiotext(e), 1983).

10   David Foster Wallace, "Up Simba," in *Consider the Lobster and Other Essays* (New York: Little, Brown & Co., 2005), 186–87.

Against sadness." The demonstration that presents itself as a cool festival overcomes such a zero degree mindset by creating Temporary Autonomous Zones that are inclusive, well beyond the multitudes of the past. This is based on a politics of affect that is purely bodily in nature and no longer fools around with 1930s visual language (as problematized by Walter Benjamin). If anything, the politics of aesthetics is audio-psychic in nature. Conjuring a sonic world that intersects with the extrasensory, politics operates within its own logic of mediality. Probes of fear, disgust, intolerance, and uncertainty are unleashed on the masses, whose emptiness is perfect for the diffusion of programs in no need of an agenda. Pure mediation of affect is all that politics requires. The job of implementation is then left up to the bureaucrats.

What are loss and desire in this digital networked age? The question may seem rhetorical, even utopian, but that's not how it is meant. Today's answer is all too often formulated in the language of offline romanticism. The way out can only be perceived as an exodus from technology as such, whereas technological proposals are often condemned as "solutionism" (Evgeny Morozov). How can we design a radical agenda that ignores both? As Chicago mayor Rahm Emanuel said: "You never want a serious crisis to go to waste. And what I mean by that is an opportunity to do things that you think you could not do before. Never allow a crisis to go to waste. They are opportunities to do big things." Let's seize that moment and become technical together. Let's forget reformist agendas that emphasize individual solutions in which participation is reduced to an input device. In fighting censorship, surveillance, and control of both states and monopolies (by dismantling actual infrastructure), there is a promise of a new culture of decentralization that will be able to negotiate its rights on a federated level with standards and protocols that benefit all.

Organizing is not mediating. If possible, we conspire offline. When we organize toward the Event, we indeed communicate, yet we do not report. This in part is what makes movements non-representable. We arrange, debate, coordinate, make to-do lists and phone calls, order necessary tools and equipment before we run outside to meet our political fate. All we need to know is this one detail: did the crowd show up? Let's keep this in mind: informing your mates is not media work. Within the logic of social media, mobilization and public relations start to blend, much to the confusion of both activists and the institutional players. Whereas for activists it is still possible to distinguish between internal channels and mainstream radio, newspapers and TV, this distinction can no longer be made clearly when it comes to the internet. Is tweeting, blogging, updating your status, posting to lists, and responding to messages symbolic work at the level of representation or a social activity?

With street protests able to scale up overnight, what we are often overlooking is the mysterious tipping point where an issue, a small network, a local controversy, suddenly flips into a mass movement. Insiders might be able to reconstruct this particular moment, but can this knowledge be translated into strategic knowledge for all? Organized networks cannot give an answer to this contemporary mystery. Whereas corporate platforms promote eternal hyper-growth through centralized sharing and updating, orgnets focus on the further development of (real-time) collaborative, decentralized networks astute to the centralizing architectures of internet communications.

Organized networks are a material condensation in software of (what Hello calls) a "commitment to commitment." Activists know that the truth cannot be found in algorithms. Models are irrelevant and are there only to administrate the world. What concepts such as orgnets do is provide coordinates for practices that structure data flows and sociality in ways that do not submit to the techniques of extraction special to social media. Comparable to the (potential) power of conceptual artworks, such proposals mingle in the titanic struggle over the planetary network architecture that defines contemporary labor and life. Code is necessary for operating systems, apps, databases, and interfaces and is highly dependent on abstract concepts. And this is where the role of software writers, philosophers, literary critics, and artists comes in. Software is not a given, an alien black box that we receive from outer space – even if we often experience it that way. It is written up by the geek next door.

Many presuppose that a "vitalist impulse" spurs the creation of networks. What needs to be questioned is the assumption that social structures arise out of the blue: there is no organic becoming without friction, only trial and error. There is a plan.[11] The aim is to reach

---

11   "We want to go beyond network-based organisation, without falling back on the model of a party. We are committed to ongoing experimentation to find the forms of collective activity needed to build a world beyond capitalism." Plan C, HTTP://WWW.WEAREPLANC.ORG/ABOUT/.

a critical mass, after which coalitions are built and separate actions snowball into a larger Event. What such Bergsonian metaphors of vitality and impulse seem to suggest is that movements in particular need to free up the energy from inside: a liberation from inner desires in order to connect to the larger, moving swarm. There is a high risk of indifference, even depoliticization that attends the vitalist impulse. What movements do most of all is surprise us (particularly those who are more intensely involved).

Instead of debunking the idea of "emergence," it is perhaps better to frame certain political configurations historically in order to get a better understanding of what could work now, in comparison to the turbulent era 40 years ago, the Golden Age of pop culture and civil unrest inclusive of social movements ranging from feminism to squatting, from armed struggle to decolonization to environmental and nuclear consciousness-raising. The biggest difference between now and then is the poor state today of the "rainbow coalition" or the patchwork of minorities, as it was once called. Instead of "affinity groups" we now have unstable networks with weak ties among their members. In today's networks we pass around status updates. If we take seriously Twitter's question, "What's happening?," the event is taken as a given.[12] The corporate assumption is that no matter how small, there is always an event in our lives that we can talk about. And it is especially the small talk that marketing experts are interested in. We neither learn of the origin of the happening nor deconstruct the urge to press the occurrence of something, somewhere, into the category of news. Is a micro-blogging service that asks the strategic question "What's to be done?" an option? What's important here is the implementation, on the software level, of a so-called "open conspiracy" (H. G. Wells). If we can accept that power is provisional, then we also need to be prepared for transformations of society that can only ever be unforeseen. And this again is why the proliferation of alternative futures, which

---

12 According to Mashable, Twitter's change of slogan in November 2009 from "What are you doing?" to "What's happening?" "acknowledges that Twitter has grown far beyond the more personal status updates it was originally envisioned to convey, and has morphed into a sort of always-on, source-agnostic information network." See Barb Dybwad, "Twitter Drops 'What Are You Doing?' Now Asks 'What's Happening?,'" November 19, 2009, HTTP://MASHABLE. COM/2009/11/19/TWITTER-WHATS-HAPPENING/.

require organizational forms, can be a centerpiece of collective design for activists, artists, architects, and educators.

There are cores, cells, and small structures that, instead of assuming the existence of a movement, work hard to get issues off the ground: groups that often barely know each other and that operate in different locations and contexts. One thing is sure: there is nothing heroic about their work. We cannot predict where their efforts will lead. Ever since the 1970s, networking these initiatives has been seen as a vital first step to get a movement up and going. This grassroots approach, so closely tied to the notion of direct democracy within a local setting, has been confronted with additional models of "summit hopping" (practiced by the Global Justice Movement), with the aim of confronting global elites that gather at G8, G20, European Union, International Monetary Fund, and World Bank meetings, and the emergence of spasmodic revolts (2011 riots in London, Paris 2005, etc.) in which networked coordination is limited during the event itself, within the existing social milieu. The wave of protests that unfolded from 2011 to 2013 can be seen as a hybrid model that contains elements of the above-mentioned forms of protest. Their issues are not (as of yet) national but are nevertheless easy to identify with on a global register. These movements scale up very fast (in part because of the use of social media during the days of mobilization), yet the crowds that gather in the street disintegrate just as quickly.

It can be productive to contrast the current debate on organization with the German "Spontis" that opposed the top-down vanguard strategies of Marxist-Leninist, Trotskyist, and Maoist groups. These days it is the masses that are spontaneous. They can no longer so easily be programmed by a consumerist-liberal consensus. However, their uprising no longer seems to be "initiated" by small groups of (anarchist) activists either. Their uprisings cannot easily be predicted or programmed. Snowballing is overtaking the capacity to organize. Ironic strategies that "blow up" conventional meaning through absurdist and playful interventions no longer seem to spark events. This is where endeavors in tactical media start to find their historical limit. The media themselves offer enough contradictory material to create dialectical cascades for a thousand and one events. But that's not the point. There is an abundance of evidence. Big data, small data: it doesn't matter. There is sufficient cynical insight. Mass consciousness is there; the work has already been done. Debates have been held and problems identified, time and again. What is lacking is the collective imagination of how to organize education, housing, communication, transport, and work in different ways.

Return of the Group: Defending Weirdness

They confiscated the passwords I would never use.
– Anonymous, 2014.

At first, social media was about our collective
unconscious. Now it's all about our collective autism.
– Max Keiser, Twitter, September 30, 2014.

In contrast to David Graeber and others associated with the North American Occupy movement, orgnets are not obsessed with the "democracy" question. The idea is not to experiment further with the community model, including with its soft governance solutions. Orgnets respond to our current moment's struggle with how technology infuses itself into the dynamics of what is considered a group. There is often the moment when the randomness of forces and events provides the necessary flip that is not reducible to the deviation capital requires in order to reproduce its power and wealth. Can there be action that is not simply interpreted as input? Revolutions will not get feedback. They just happen. Organization after revolution cannot rest comfortably with reproducing the party form or assuming that general assemblies are the utopia of participatory politics, as Jodi Dean and others would have it.[13]

Organized networks start with refusal, which includes the exodus from participation. But withdrawal or inactivity is also not the solution and is often a disastrous trap that leads to depression and incapacity – both of which are products of what Franco "Bifo" Berardi and Bernard Stiegler call pharmacological capitalism.[14] There Is An

---

13   Jodi Dean, *The Communist Horizon* (New York: Verso, 2012).

14   See Franco "Bifo" Berardi, *Precarious Rhapsody: Semiocapitalism and the Pathologies of the Post-Alpha Generation*, trans. Arianna Bove, Erik Empson, Michael Goddard, Giusppina Mecchia, Antonella Schintu, and Steve Wright (London and New York: Minor Compositions, 2009). See also Bernard Stiegler, *Taking Care of Youth and the Generations*, trans. Stephen Barkerc (Stanford: Stanford University Press, 2010) and *What Makes Life Worth Living: On Pharmacology*, trans Daniel Ross (Cambridge: Polity, 2010).

Alternative. But the paradox is that all the alternatives are largely already known. We know how to address the eco-disaster problem, how to deal with child care, health care, and schooling. Alternative medicines are available, as are models and practices for bio-food, urban farming, recycling, and free software. The list goes on, and the alternatives work. It is no longer the point that alternatives are ill conceived or bugged with communist dogma. Organized networks do not emerge to prove that alternatives are viable and work. We know that many of these alternatives can be fed into and complement the capitalist system. Just look at the carbon trading economy for an extreme example.

Organization beyond exodus entails the work of social design, of collective experiences free of disciplinary measures that produce new restraints. Organizing networks encompass the invention of new protocols and standards of connection, systems of finance, and cultures of communication. Yet the question of social form remains. What does it mean these days to join a group? First, there's the social affiliation and desire for relation. The group provides a new identity for participants and the patchwork of sociality. Membership has become a compulsory social activity where commitment itself expires as the individual grazes from one cause to another. Groups therefore are clearly transient forms of experience and activity. This is a problem that organized networks share. But unlike groups the transient character of organized networks is, in fact, their primary condition of sustainability through the serial accumulation of experience and collective production of knowledge. The group form obviously persists, but the intentionality of belonging disappears.

While the group was remarkable for the way in which it proliferated across Western countries in the 60s and 70s, giving rise to new social movements, its borders were always very contained, with limited numbers of individuals sharing the passage of experience through life.[15] The group sustains its own rationale for existence, developing its own measures of success and achievement, becoming only more intolerant of deviance and difference. The network does not have this feature. The group may be wrought with psychodrama and a tendency toward excessive introspection, but at least it's a narrative that does the work of self-affirmation. Again, networks do not have this feature.

---

15 See Mary McCarthy, *The Group* (London: Weidenfeld and Nicolson, 1963).

Yet, for all this, even the group could not sustain itself. It was a generational phenomenon and had disintegrated by the neoliberal 90s. And this is part of the problem for new social movements today, which cannot depend on these groups – either the core of the movement has gone, or the organizational vanguard have become unobtainable as groups transformed into fully professional entities complete with lobbyists, fund raising, corporate codes of conduct, and so on.

Networks have a hard time explaining themselves and producing a story of their own existence and becoming. So why are narrative and network seemingly incommensurable? There's the obvious media answer at the level of technical composition. Scholars of the novel, film, and television have extensively studied the story of the group. There is a fascination, even an obsession, in these media with the element of exclusion inherent to the dynamics of the group, even if this is not always foregrounded. The group appeals to a stable set of social codes and rules. Today, the resurgence of interest in the group in social life across Europe's cities arises at a time of increasing pressures on urban infrastructures, social space, and employment. Like urban "tribes" and gangs, the group functions as a border technology against unwanted intrusions. It holds a neo-traditionalist element these days, a mourning for the passing golden era of Fordism and its temporary stability against which the group could rebel. In this network era there's very little sign of sectarian groups – hate groups and pedophile rings have learned about the dangers of the network, so they revive the group within careful firewalls and encryption. Is there a lesson here for movements whose default technologies of organization are Facebook and Twitter?

Organized networks reverse the social procedures of the group in that they start from the network level. The network is not a voluntary, bottom-up association of groups (as in the 70s ideology) but manifests itself as the dominant everyday form of social life. The organized network is our basic unit. How orgnets relate to each other, and to other organizational forms, is an interesting question but one that cannot properly be addressed as long as the contours of orgnets remain only barely visible. We have to get a better understanding of what's going on (and what's possible) at this basic level. Graeber writes: "It's always better, if possible, to make decisions in smaller groups: working groups, affinity groups, collectives. Initiative should rise from below. One should not feel one needs authorization from anyone, even the

General Assembly (which is everyone), unless it would be in some way harmful to proceed without."[16] From the network, the group may emerge. And from the group, the network becomes organized.

In the former West, the crowd is no longer a threat. At best it is a carnivalesque symbol, a signal for frictions in society (but which ones?). We are all products of *The Century of the Self* (Adam Curtis, 2002). No matter how powerful the image of a large gathering of protesters may be, there is always an element of entertainment in it, produced for individual consumption. For many, rioting is a form of extreme sports.

How to create non-eventful forms of organization? If "crystals" (Sponti, Leninist or otherwise) are no longer the cause of events, we might have to move away from cause-effect thinking altogether. The Spectacle, with its auto-generated intensity of affect, goes against the Time of Organization. The complex and muddy coordination between different levels and interests cannot beat the real-time spread of memes. Organized networks grow in response to the universal solution of the algorithm. We organize against aggregation, multiplication, and scale. We want seriality, not scale, and voluntarily step back from the viral model that inevitably culminates in the backlash of the IPO (the Termidor of the dotcom age). When a company decides it is "going public" and is taken to the stock market, it's not too long before we see the evisceration of everything that defined that organization as unique and appealing. This includes management buyouts and the first wave of take-overs. How many fallouts can you afford before you no longer have any friends?

One issue that is difficult for orgnets to address and resolve in some functional way is the question of leadership. The celebrity SPO (Single Person Organization) model, as practiced by WikiLeaks' Julian Assange, has proven to be disastrous. So too is the closed "cabal" circle that runs Wikipedia. Less controversial is the rotation of moderators in the case of some email lists (such as Empyre) and the "Karma" voting systems (as in the case of Slashdot) that are operated by users. News aggregator Hacker News uses a similar system to weigh the ranking of geeky forum websites.

The other hot item is the issue of political demands: who formulates them and decides whether they are necessary in the first place? It is the enunciation of the demand that compels networks, media, and the representative bodies of organizations with something at stake to meet and devise plans. Recall the orchestration of select media organizations by Julian Assange in preparing the Afghanistan classified cables and the follow-up meetings after their publication between government authorities, law firms, media advisers, and any number of self-interested individuals and entities. Whatever position you may care to take on the jaded spectacle of Assange, we can nonetheless diagnose the force of its event as that which is precipitated by the instantiation of the demand. This also reminds us to take care in separating expression (the demand) from the logic of eventism (the spectacle). Organizing networks across iterative processes of seriality builds the enunciation of the demand within a logic of modularity. The recombinatory effect of seriality guards against the demand being attributed to any single issue or agent, and explains why the demand remains a useful strategy in displacing networks from the all-too-common and politically disastrous tendency toward self-affirmation. When that happens, it's game over.

In the current technological landscape, social media do not focus on transparent technologies of agenda-setting, discursive preparations of policies, and decision-making procedures. This is in the interest of neither the news and entertainment industries nor marketing. Graeber writes: "We have little idea what sort of organizations, or for that matter, technologies, would emerge if free people were unfettered to use their imagination to actually solve collective problems rather than to make them worse." It is clear that this issue can't be limited to the media diversity issue and real existing differences of interest. Relevant for both social movements and organized networks is the (potential) influence of the Online Absent Other. It is both politically correct and a comfort for all to state that others who cannot be there in person will be able to steer and not merely witness events. But the discourse of networked forms of participatory politics has its limits. "Being so connected to something you are disconnected from is, I believe, deeply disturbing to your psyche. Sooner or later things make sense and your mind realizes it's been seeing and reading one thing and living another. At that moment it just happens – you 'go dark.' Vanish."[17] Organizing networks is a practice of orientation that will

---

17   Cameran Ashraf, "The Psychological Strains of Digital Activism," Global Voices

always require the full – and often deeply unwanted – disturbance of the senses. Only then will the bodies that matter issue demands that count (#Ferguson).

Advocacy, posted on April 17, 2013, HTTP://ADVOCACY.GLOBALVOICESONLINE. ORG/2013/04/17/THE-PSYCHOLOGICAL-STRAINS-OF-DIGITAL-ACTIVISM/.

10.

ARCHITECTURES OF DECISION

Far more dreadful are social milieus, with their supple
texture, their gossip, and their informal hierarchies.
– The Invisible Committee, *The Coming Insurrection*,
2009.

Please do not share this announcement with any
journalists. I have selected your profile as a trusted node
of connection in the cultural professional networks and
would hereby like to invite you to access a platform of
pirated daily financial news.
– Confidential announcement, 2014.

THE DIVERSITY OF THE MOVEMENT, THE INFORMALITY AND SPEED OF
the network, the rituals of the assembly, and the formal power of
the party: each political form has its distinct features and dynamics.

Medium theory always taught us that expression is shaped by the contours and material properties of communication technologies. The same can be said for these organizational political forms. They hold a capacity to mediate and conform to an extent dictated by their typology, enabling certain processes while frustrating others. No matter their internal variation – there are many different kinds of networks, just as there are assemblies and so forth – there is something distinct about their organizational forms.

How to comprehend the emergence of large global protests and the rise of networked movements? The concept of organized networks can neither explain such phenomena, nor is it a response to them. First and foremost, organized networks arise from the growing discontent with social media and their presumed role as the motor behind the current popularity of protest. But perhaps more importantly, orgnets are a response to the problem of organization and institutional form, and have to be understood as a response to the current contradictions. The abstraction of democracy is not so often a motivating force that brings bodies and brains to the street. The 2014 Hong Kong uprising is a clear case in point, where the experience of economic misery projected into the future was a powerful enough catalyst for political action and the production of subjectivity.

Four primary features define the current situation of communication systems for orgnets. First, short-term communication that evaporates after resolving task-based organization. Second, mobilization that is focused on connecting core organizers with politicized masses. Third, the facilitation of deliberation, discussion, and debate. And fourth, processes of decision making that demonstrate populist governance at work (e.g., assemblies). The concept and practice of organized networks is only one of a whole range of possibilities circulating in the field of political design.[1] Our thesis is that the current wave of protests is key to

---

1   Possible solutions would require a mix between offline practices such as the
      assembly (as discussed by David Graeber in *The Democracy Project: A History, a
      Crisis, a Movement* (2013) and Marina Sitrin and Dario Azzellini in their book
      They Can't Represent Us! Reinventing Democracy from Greece to Occupy (2014)
      and technical online experiments such as the LiquidFeedback decision-making
      software (see HTTP://LIQUIDFEEDBACK.ORG/ and the work of Anja Adler from
      the NRW School of Governance in Duisburg, Germany), or the Loomio de-
      cision-making software which assists groups in collaborative decision-making

the production of new forms of organization beyond traditional forms such as trade unions, tribes, and even social movements.

The Network Decision

The question of decision is one we see as central to the organizational form of networks. For Jodi Dean, this is the function of the party. "The primary organizational question, then, is what might a party look like for us? What features might install in it the necessary discipline, flexibility, and consistency necessary for building communist power?"[2] The assembly attempts to do what Dean desires for the party, namely the instantiation of the moment of decision making through consensus building. As a form of direct democracy, people try to go beyond the particular – single issues, the core topic, the fragmentation of desire, will, cause, and so forth. The assembly form tries to sideline mediation. There is an implicit critique of social media here, and a desire for the corporeal as the pure scene of social relations.

For Dean and the Communist Co. (Žižek, Badiou, and their droves of disciples), democracy is a messianic moment, driven by a passionate urgency, undiluted and unadulterated by the technical, which is about exclusive knowledge (engineers) and political economy (the dirty corporation driven by commercial interests). This sentiment is shared by the "assembly" advocates, even though the political form of the assembly is considerably more experimental. It operates according to the logic and immediacy of the event, while the party is the established organizational structure that has proven its form as a device for deliberation and organized power throughout modern history.

This is no less the case today, as the political form of the party adapts its operations, techniques of reproduction, and processes of decision making to the metric-driven experimentation of data analytics, enabling a real-time modulation of the relation between parties and populations.[3] As noted by Zeynep Tufekci, technical infrastructures

---

processes in an attempt to overcome the short lifespan of social phenomena such as swarms and smart mobs.

2   Jodi Dean, "The Question of Organization," *South Atlantic Quarterly* 113.4 (2014): 834.

3   See, for example, Zeynep Tufekci, "Engineering the Public: Big Data, Surveillance and Computational Politics," *First Monday* 19.7 (2014), HTTP://FIRSTMONDAY.ORG/OJS/INDEX.PHP/FM/ARTICLE/VIEW/4901/4097.

consisting of "blogs, micro-blogs and online social media and social networking platforms" foreground a new system of algorithmic governance overseen by a "small cadre of technical professionals."[4] The technical expertise and algorithmic operations special to such modes of governance results in a party-form whose machinations are even more obscure, abstract, and removed from processes of accountability. Algorithmic governance further dislocates any public comprehension of how policy making, for example, is developed out of a relation with the supposed empirical qualities of lived experience. Metrics, and what architectural theorist Reinhold Martin terms a "numerical imaginary," function as the new mediators of our machinic relation to material phenomena and the modulation of desire.[5]

The act of decision is determined by the politics of parameters, whose expression as architecture in the form of algorithmic operations establishes a correlation between the party as an organizational entity and other systems dependent on complex computational procedures, such as high-frequency trading. A paradox emerges in which data analytics makes possible the reinvention of the party form within a paradigm of social network media, while also serving to effectively deontologize twentieth-century organizational forms such as the political party and the trade union in such a way that makes its decision-making processes indistinct from any number of other social, economic or political system driven by algorithmic operations. The sovereignty of the algorithm, in other words, renders the borders and functional capacity of organizational forms in ways that are indifferent to an ontology of the visible. This is why a correlation can be made between what otherwise seem entirely different, even incompatible, organizations (as operational systems). Organizational forms whose acts of decision have not yet been captured by the power of the algorithm and data analytics remain a social-political force to be reckoned with. An ontology prevails, one that is defined by the unruly and conflictual relation between the social and the technical.

For venture capitalist Peter Thiel, the PayPal founder who is on the board of directors of Facebook, the social is a conspiracy of invisible

---

4   Ibid.

5   Reinhold Martin, *Mediators: Aesthetics, Politics and the City* (Minneapolis: University of Minnesota Press, 2014), 1.

bonds, of old boys networks.[6] It is not sufficient for an organization to have good people, the best in their field, and so on. The assumption that tribal bonds of efficiency result in strength and organizational power cannot be guaranteed by assembling the leading experts in one room. This lesson was made clear in the film adaptation of Michael Lewis's *Moneyball* (2011) with Brad Pitt and Philip Seymour Hoffman. In that story, the best possible baseball team was assembled not through the acquisition of star players, but rather though the algorithmic selection of a team whose aggregate performance complemented and transcended what were otherwise assumed to be individual weaknesses. The lesson here for organized networks is to build on that totality of strong ties that advance a particular cause or movement and not succumb to the allure of diversification and the dilution of talents that all too often defines the democratic gesture. When manifest in the party, such modes of organization stagnate with a cohort of brilliance whose individualized desolation is unable to commit to the collective decision of the project or movement.

In this age of social network media, it really isn't conceivable any longer for the party to emerge from within itself. In other words, the plebiscitary of the networked conversations is the dominant social-technical form, which means the party can only emerge from within the culture of networks. As Rodrigo Nunes writes: "*Even if* a return to the party-form were found to be the solution, the party would no doubt have to emerge from existing networks."[7] A populist party such as Podemos, for example, will always carry the trace of the network in its myth of origin: the movement of the squares (M15) serves as a core component of its organizational ontology, despite its image of an electoral war machine under the leadership of Pablo Iglesias. The network logic has redefined the delicate balance of unity and diversity. That's the lesson learned: irrespective of where you want to end up as far as your preferred organizational form goes, in order to know your object you will necessarily start to diagnose your network.

---

6   Peter Thiel with Blake Masters, *Zero to One: Notes on Startups, or, How to Build the Future* (London: Penguin Random House, 2014).

7   Rodrigo Nunes, *Organisation of the Organisationless: Collective Action after Networks* (Leuphana: Mute and Post-Media Lab, 2014), 11, HTTP://WWW. METAMUTE.ORG/SITES/WWW.METAMUTE.ORG/FILES/PML/ORGANISATION-OF-THE-ORGANISATIONLESS.PDF.

There are two dimensions to this: the technical and the social. All too often the technical remains insufficiently addressed and too easily dismissed. The social, on the other hand, is privileged as the primary logic of networks – at least within social media where the precept of interface design is to accumulate more friends whose data entrails feed an economy of extraction and recombination.

In the case of networks there are, broadly speaking, social media platforms that are closed proprietary worlds at the technical level. The attempts to generate alternatives have to date been limited. Diaspora was a case of an open source network that entered campaign mode dependent on crowd-sourced funding, short mobilizations, and global protests to garner mass attention with the hope of signing up enough users to make an operational claim as a viable alternative to commercial providers.[8] The difficulty for Diaspora has been that it could not scale up and become adopted by a movement galvanized by the event. It also wasn't sufficient at a design and technical level to copy or mimic the juggernaut of Facebook.[9]

The scalar strategy has to be questioned because such old broadcasting tactics all too often lead to weak ties with little impact. We know how to scale and the world of PR and advertising does it all the time. So do orgnets (as concept and practice) become a form of retreat or an elite avant-garde strategy to prepare and anticipate change and future implementation? Do they offer new forms of social interaction? Is it ridiculous to think of a Bauhaus of net-culture that provides the blueprint for future mass production and distribution? The network protocols have a similar and bigger impact – more general than IKEA. They are not a lifestyle. There is currently no choice in the adoption of media architectures.

Post-Snowden and the PRISM revelations, social media software is tasked with the additional demand to ensure encryption. The March 2014 hype around the ad-free social network Ello showed there was a great need for an alternative to the Facebook and Twitter monopoly.[10]

---

8   Diaspora: HTTPS://JOINDIASPORA.COM/.

9   Since 2011 the Unlike Us network has been collecting and discussing experiences with alternative social media platforms. Besides three conferences and a reader, this has mainly been done on the Unlike Us email list: HTTP://NETWORKCULTURES.ORG/UNLIKEUS/.

10   Ello, HTTPS://ELLO.CO/BETA-PUBLIC-PROFILES.

The problem here was that, as often is the case, the platform launched too early and became overhyped through the traditional media. Because of the premature release the developers hadn't built a solid foundation of early users and adopters. The larger issue here for developers of network software is the pressure from users for a post-PRISM alternative, resulting in insufficient time for developers to fine tune the software. The demand for instantaneous solutions has resulted in half-baked, bug-ridden products being unleashed with a near certainty to fail, crippling many efforts to build new networks.

We also need to address developments coming out of the right-wing, techno-libertarian start-up movements. Many of these proponents hold ideas and proposals not so different from those of the open source movement, academic left, and even the art world. The call, for example, by the conservative tech revolutionary Peter Thiel to dismantle formal university training and college education is one that chimes frequently within groups gathering around free and public universities. Thiel prefers the commitment-to-cause inside cults to the nihilism and non-engagement of the consultancy class. He proposes to "take cultures of extreme dedication seriously. Is a lukewarm attitude to one's work a sign of mental health? Is a merely professional attitude the only sane approach?[11]"

While the sentiment may share a similar distaste for state power and the society of crippling debt – manifest also in the "sharing" economy such as Airbnb and Uber, which strive to promote services unshackled from the regulatory control of the state – the political agenda of the start-up world is often obscure, presenting itself as a space of neutral tools for free.[12] But this is far from the case, and all too often radical left-wing movements sign up unwittingly to a techno-political ideology of libertarianism that is, if nothing else,

---

11 Thiel, 124.

12 See the extraordinary ongoing reporting by Paul Carr for PandoDaily (HTTP://PANDO.COM/) on Uber and the more general analysis of Sebastian Olma, "Never Mind the Sharing Economy, Here's Platform Capitalism," October 16, 2014, HTTP://NETWORKCULTURES.ORG/MYCREATIVITY/2014/10/16/NEVER-MIND-THE-SHARING-ECONOMY-HERES-PLATFORM-CAPITALISM/ and Trebor Scholz, "The Politics of the Sharing Economy," May 19, 2014, HTTP://COLLECTIVATE.NET/JOURNALISMS/2014/5/19/THE-POLITICS-OF-THE-SHARING-ECONOMY.HTML.

massively contradicting the political sentiments of movements. This is because the movements are largely disconnected from development and consider technology simply as a functionary tool when in fact it profoundly shapes the production of subjectivity and instils practices with political values that are far from neutral. This only becomes really visible on the surface in a city such as San Francisco where these political tendencies collide, with activists attacking the Google bus that transports tech-designers in the secure world of a charter bus.

The Invisible Organization

The invisible is not the virtual. Thanks to the Snowden revelations, we continue to live through the trauma of materiality accompanying the seemingly invisible realm of what was fondly referred to in the 1990s as "cyberspace." The utopian spark of the Hong Kong protests and occupation brought off-the-grid organization to widespread attention. While this practice ensured a mode of connection not dependent on commercial infrastructures, it was not without implications for the sustainability of orgnets. Blue-tooth communications using applications such as FireChat are inherently insecure and indeed invite the enemy to infiltrate what might otherwise have been secret planning sessions.[13]

The other aspect of off-the-grid computing concerns the relation between memory, archives, and the sustainability of movements. Without a common repository, off-the-grid computing lends itself to the task of organization but not the storage of collective history. One exception can be found in the collective archives of political dissent such as MayDay Rooms, which exemplify the work of collaborative constitution and the mediation of memory for social-political movements.[14] Moreover, projects such as these acknowledge the social-technical logics of retrieval: the power not of the net and its infrastructures but of the network-form itself.

---

13   See, for example, Noam Chen, "Hong Kong Protests Propel FireChat Phone-to-Phone App," *The New York Times*, October 5, 2014, HTTP:// WWW.NYTIMES.COM/2014/10/06/TECHNOLOGY/HONG-KONG-PROTESTS-PROPEL-A-PHONE-TO-PHONE-APP-.HTML and Archie Bland, "FireChat: The Messaging App that's Powering the Hong Kong Protests," *The Guardian*, September 29, 2014, HTTP://WWW.THEGUARDIAN.COM/WORLD/2014/SEP/29/ FIRECHAT-MESSAGING-APP-POWERING-HONG-KONG-PROTESTS.

14   MayDay Rooms, HTTP://MAYDAYROOMS.ORG.

Short encryption services that operate like a secure free local SMS used by thousands on a particular location accommodate the rumor-like forwarding of event announcements and actions. But is this sufficient to the task of organization? This indeed was one limit-horizon experienced by core organizers in recent uprisings the world over. Again, we refer to the political demonstrations in Hong Kong in 2014, which were mobilized through off-the-grid computing and the use of a Bluetooth app called FireChat. Here, the use of an open access system meant authorities were just as informed as movements. And without the technical capacity of archiving communication, the Hong Kong movements leave open the question of how to galvanize future movements and constitute subjectivities without reference to an archive of dissent.

Does the invisible organization have an invisible committee? Occupy circles do not like to talk about the informal groups behind their spectacle-type assembly meetings. And it's not hip to discuss who's in charge of the agenda-setting of their consensus theatre productions. Orgnets? In the case of classic guerrilla organization as depicted in Gillo Pontecorvo's film *The Battle of Algiers* (1966), anti-colonial insurgents are organized across a network of cells whose relation to each other are structured according to a triangular logic, thus protecting but also obscuring the central management. This social form of invisible organization is not readily duplicated within the logic of networks whose technical parameters may section off core organizers and administrators from members and participants, but does not guarantee the sort of structure of security and invisibility afforded by pre-digital techniques of social-political organization. Outside of encryption, the digital at the level of everyday popular use is a technology of transparency. Yet at the level of infrastructure, the digital is more often enclosed within the black box of proprietary regimes and heavily securitized data centers.

This prompts the question of alternative (tech) infrastructures and the need to share expertise about what works and what doesn't (how to deal with trolls and spies, domain name disputes, email overload, at what point switch to crypto, etc.). The future of organization in a post-Snowden landscape requires the generation of new protocols and perhaps off-the-grid computing made secure. To create new protocols within the sort of techno-political environment just described, however, is another task altogether. Part of this work requires attention to devising strategies and tactics that facilitate and sustain political intervention over time. But crucially, such work must also address the question of organization coupled with technical and infrastructural issues related to social media use, something we see missing from the sort of campaign handbooks written by London School of Economics civil liberties activist Simon Davies.[15] His campaign ideas fail to register the political stakes of organization among non-citizens who may reside in the territorial borders of the nation-state, but hold none of the privileges accorded to the citizen-subject. Moreover, his toolkit of strategies and tactics for campaigning is wedded to a media logic of representation. He has nothing to say about how media forms and organizational networks play a core role in the sort of campaigning he advocates. It's almost like we never left the broadcast age and its conspiratorial moves behind the scenes, aimed to "manufacture public opinion." These days "ideas for change" should explicitly deal with distributed nature of both organizations and mass communication. The internet is not a black box anymore.

German media archaeologist Siegfried Zielinski does not believe in the "protocol" approach that locates the center of power in code, while turning this insight into a political strategy. This is the core of the nerdist philosophy: it is neither content nor interface that determine our situation. Zielinski: ". . . I acknowledge my powerlessness. The position from which I believe it is still or is again possible to formulate criticism is located on the periphery, not in the center."[16] We do not believe that either the nerd or the artist will enlighten us with the Truth. Key for us is not the perpetuation of tinkering on the margins of obscurity but in fact focusing very clearly on the network architectures at the center of power. For Zielinski, there is a political periphery of informality dedicated to art initiatives, counter-designs, alternative interfaces, and so on. However, we take our cue from the various companies such as Red Hat, Canonical, and SUSE involved in the Linux enterprise who, among others, located their businesses in the center of technological development in Silicon Valley despite the

---

15 Simon Davies, *Ideas for Change: Campaign Principles that Shift the World*, December 2014, HTTP://WWW.PRIVACYSURGEON.ORG/RESOURCES/IDEAS-FOR-CHANGE.

16 Siegfried Zielinksi, *[...After the Media]*, trans. Gloria Custance (Minneapolis: Univocal Publishing, 2013), 21.

general Linux radical agenda of a free, open, and distributed communication architecture.

The design of alternative protocols must first reckon with the architecture of communication and control. This political economic geography of centers, not margins, extends to the infrastructure of storage, transmission, and processing: namely the highly securitized, hidden territory of data centers (also known as server farms or colocation centers). Until we know more about the technical operations, communication protocols, legal regimes, design principles, and social-economic impact of such infrastructure, the capacity of movements to make informed decisions about how to organize in ways that both support and secure their interests and agendas will remain severely circumscribed. Orgnets may have an important role to play here in terms of coordinating collective endeavors of critical research into such infrastructures of power.

Introducing the Post-Digital Organization

Organized networks always take place in informational settings and this is why the post-digital becomes a relevant topic and condition to address. Questions of organization and "political design" these days cannot be separated anymore from the IT realm. The concerns of the informational carry over. This is not good news for the development of organized networks, where so many of the alternatives in the making are abandoned as the supremacy of templates takes command. Post-digital organization in such a context is also then about knowledge organization. The university has been an institution whose autonomy is undermined, in question, and so forth and this is only made clearer in a post-Snowden context.

According to Rotterdam media theorist Florian Cramer, "'Post-digital' first of all describes any media aesthetics leaving behind those clean high tech and high fidelity connotations."[17] It is no longer the remit of engineers. Post-digital = Post-digitization. The phase of digitization is complete as a process of implementation and integration. We are no longer speculating about the arrival of the new. The systems are in place. Our task is now to map its impact, in real-time: instant theory. When the scanning is done and records logged, meta-tagged,

---

17   Florian Cramer, 'What is Post-Digital?', *A Peer Reviewed Journal About // Post-Digital Research* 3.1 (2014), HTTP://WWW.APRJA.NET/?P=1318.

and uploaded in the database, the bureaucratic end of digitization takes command. The transition is over and the story can begin in which the hidden dream of a post-digital renaissance where the old values of humanistic inquiry would resurface and once again supply society with the moral compass and grand stories through which to conduct life.

The post-digital is not about less digital, it's about the digital pushed to the background (Cramer), in part because it is made invisible and integrated into everyday life, but also because we've mastered it (or should have). The digital was never about the digital precisely because no one ever knew what was in the background. It just worked as general infrastructure in much the same way that electricity does – the magical cyberspace that was connected. The post-digital therefore needs to be understood as a process of demystification – part of the *Entzauberung der Welt* (Schiller). Disenchantment does not capture the mythical element. There is a lyrical element that is erased in the process of normalization and the rise of technocratic culture and the administrative world (*die verwaltete Welt*).[18] And here, we can say that the post-digital is coincident with the logistical worlds of global supply chains because they too are very much about calculating activity and the repetition of movement within enterprise architectures with the aim of extracting value from the accumulation, analysis, and commercial sale of data.

We can note the place of digital humanities in this phase of digitization where much of the debate has now moved on to data analytics, visualization, and the development of dynamic research methods. Amidst issues around data securitization are a concern for institutions, including those engaged in higher education, hosting these projects not only for reasons of sound ethical practice with regard to data management. Of perhaps greater concern is the potential market value and consequent proprietarization of data generated within quasi-public institutions where the logic of data is increasingly understood in economistic terms. The lesson of social media, in other words, has migrated over to the disciplines and the production of knowledge. We can only sense what platform capitalism will have to offer us in the near future in terms of a brutal yet invisible destruction of the social.

For Cramer, the new normality of the post-digital corresponds with ordinary, everyday use of template software. The post-digital sits

---

18   See HTTPS://DE.WIKIPEDIA.ORG/WIKI/VERWALTETE_WELT.

comfortably with off-the-shelf aesthetics and pre-formatted cultural expression. Such a condition frequently carries over to the organization of networks, where social-technical non-dynamics are the default mode of coordination and communication. Template culture for political movements results in a certain indistinction across what are otherwise enormously varied social collectives and movements who are responding quite often to very particular social and political issues. The template mode of organization shifts authority in some respects from the engineering class to the networked multitudes who need to self-administrate their dreams in excel sheets.

The roaring nineties were bedazzled by the allure of shiny interfaces of multimedia projects, even if their execution was frequently short of the mark and poor at the level of content. Within the post-digital scene, aesthetics is drifting off to the "maker" culture, which is notable for two key developments. First, is a media aesthetics of nostalgia for a 1980s DIY media culture. And this taps into the second prevailing feature of the post-digital, namely an experimental interface aesthetics that bridges the material with the digital and creates new one-off hybrids ("new aesthetics").[19]

The post-digital is beyond the dialectics of old and new media, since with digitization all communication forms are computational in nature (appearances might be different). We can definitely say the digital has been enormously powerful at the level of the imaginary. The post-digital therefore implies a time and space within which new imaginaries can emerge that are not beholden to the mysteries of engineers or the obscurity of internet governance circles and geek enclaves such as GitHub. So a politics of the post-digital would be one that confronts and makes visible the submersion of communication into the vaults of secret data centers scattered about the globe. Rational technocratic organizational culture can only be more horrendous when it disappears as an object of fear and force of control. This is not to get conspiratorial and whatnot, but rather to know how communication, culture, and economy are operative in a digital world. Otherwise it's the story of a great somatic slumber.

We have often asked how something comes into being as a singular event. The question should now focus on the materiality of the

---

19   See David M. Berry and Michael Dieter (eds), *Postdigital Aesthetics: Art, Computation and Design* (London: Palgrave Macmillan, 2015).

continuum. Media and cultural studies can only study an object or phenomenon once it's there. But today that's not enough if you are always running behind the facts. This is the weak point. The alternative is not predictive analytics of big data. Pattern recognition supposes equivalence between data, its algorithmic organization, and external referents within the world. And even though there's increasingly an integration of data, technology, and life (Internet of Things), it is at least for now not totalizing. Life still escapes. So while we can say media are constitutive of the experience of communication and even produce unforeseen effects, there is an outside to media. And a post-digital theory of media therefore has to know not just what it incorporates (as in the measuring of systems), but it also needs to know that which is external to it since these will be new spaces of capture in the design of media architectures that drive capitalism.

The post-digital is a form of submission. The ubiquity of the digital, its thorough integration into the routines of daily life, signals acquiescence to the hegemony of standards and protocols. In its generalization, the post-digital withdraws into the background. We are no longer conscious, or even less conscious than say we were during the 1990s, of the architectures which support communication and practice. Often the presence of technology would be sufficient to prompt an address. But with the slide into the background of sensation, we have no clue what constitutes the media as an object of communication, which operates on a spectrum outside human cognition and perception. The rise of big data registers the limits of human cognizance when set against computational power.

Can we say there is equivalence between the post-digital and the post-human? For Nicholas Carr, human agency prevails as the invisible remainder following the automation of economy and society.[20] The human is the last resort and upholder of an ethical existence and the necessary component for the system to survive. Since Taylorism and Fordism, the human was consigned to a future of redundancy and eventual obliteration. This scenario has played out within a networked, informational paradigm through the figure of "free labor," where value is extracted from social relations made possible by digital media of connection. Once this horizon is eclipsed, which we start to

---

20 Nicholas Carr, *The Glass Cage: Automation and Us* (New York: W.W. Norton, 2014).

see with the rise of the quantified self movement, there is no resource left for capital accumulation, since time also has been effectively eradicated with the nano-speed of high frequency trading, which generates something in the vicinity of 80 percent of global financial transactions. At the end of the day, the inert body of the human persists and retains a capacity for thought and action, which include importantly for techno-capitalism the power to decide, which is a power to distinguish. Paradoxically, the loss of agency through automation conditions the vitality of what remains of the human.

New schools of literacy for the invisible will emerge to address the power of automation and the post-digital. The Snowden effect is just one index of a nascent awareness of technologies of capture supported by the selective openness of enormous datasets collected on behalf of the dominant IT companies for the techno-surveillance complex. To shift your operations to less noticeable IT providers is of course a false security, since they are harnessed to the infrastructure of power such as data centers.

To summarize, we propose to shift attention away from mobilization and event making to collaboration in order to provide rapidly emerging social movements and "global uprisings" with more sustainable organizational tools. This includes producing concepts that correspond with the social-technical dynamics of practice, and which operate as an architecture through which things get done. The social ties within protests are tightened through the work of organizing networks. Focusing on the consensus spectacle of the assembly or the nostalgic return to the form of the political party is fine if you are seeking distraction *because movements are temporary and cannot make decisions.*

11.

MOVEMENTS

All revolutions are impossible until they happen. Then
they become inevitable.
– Albie Sachs

CAUGHT IN THE REAL-TIME REGIME ALL WE CAN DO IS SPECULATE
about the future value of concepts. Over the past decade we've worked
together on many texts producing one concept: organized networks.
Prior to examining the challenge of organization, let's examine the ups
and downs of the network paradigm. Organizing presupposes a Will
to Act. So, before we regress into the "interpassivity" that dominates
our conspiracy age, it's important to address the status of the online
self: how can we prevent portraying ourselves as victims of fake news?
How will we recover from the Big Data regression? If refusal of social
media is no longer an option, how do we master the fear of missing
out and take matters into our own hands?

It was in the year 2016 that networks were pushed aside by the
overarching term "platform." This unnoticed shift, not just in the literature but also in the common language, was reflected in two rather
different publications: Benjamin Bratton's grand design theory, *The*

Stack: On Software and Sovereignty, and Nick Srnicek's critical essay, *Platform Capitalism.*[1] Both frequently use the term "network" but no longer give it much significance.[2] Such a tendency is in line with American business literature on the topic, such as *Platform Revolution* by Geoffrey Parker and others.[3] In the age of Uber, Airbnb, Google, Amazon, and Facebook, networks have been downgraded to a secondary organization level, a (local) ecology, only significant for user experience. It no longer matters whether the network as a (visualized) set of correlations has any meaning. Networks can be big or small, distributed or scale-free. As long as their data and potential surplus value can be exploited, everything runs smoothly.

What can we hold up against the nihilist reality, assuming we want to "come together"? One proposal would be to de-historicize and re-design the media-network-platform triangle into layers – or stacks, if you like. But platforms are not our destiny. Let's sabotage Kevin Kelly's notion of "the inevitable."[4] In the same way as media are not merely about communication, networks are more than social media. How can we upset the Hegelian synthesis that is presented to us as the best of all possible worlds? How can we undermine the logic of prediction and pre-emption. What does it take to disrupt the correlation machines?

The much desired "commons" will not be offered up to us on a plate. We need to get our hands dirty by collectively building up old school independent infrastructures before we can begin a detox program that cleaves us from our dependency on "free" services. How can we collectively design "commoning" as if it were a popular sport? To develop cooperative alternatives to the data center logic of Silicon Valley and East Asia's "smart cities" is not a mystery. What could be

---

1   See Benjamin Bratton, *The Stack: On Software and Sovereignty* (Cambridge, Mass.: MIT Press, 2016) and Nick Srnicek, *Platform Capitalism* (Cambridge: Polity Press, 2017).

2   If there's something like Dark Deleuze (Andrew Culp), how long should we wait for Critical Castells? See Andrew Culp, *Dark Deleuze* (Minneapolis: University of Minnesota Press, 2016).

3   Geoffrey G. Parker, Marshall W. Van Alstyne, and Sangeet Paul Choudary, *Platform Revolution* (New York: W. H. Norton, 2016).

4   See Kevin Kelly, *The Inevitable: Understanding the 12 Technological Forces that Will Shape Our Future* (New York: Viking, 2016).

today's equivalent of the "temporary autonomous zone"? If once there was a fear of appropriation, these days there is simply no more time and space where subversion can unfold. What's needed is a new form of shadow, since we can no longer hide in the light of consumer culture and pop aesthetics (Hebdige).[5] Once the "meme" has been designed, there are enough real-time amplification channels available to spread the message.

Looking Back at Network Cultures

The historical question we need to ask here is why networking became such a big topic in the first place – and what this could teach us, a good decade later. The trouble may all have started with the introduction of the "scale-free network." With the dramatic drop in the prices of hardware, software, and connectivity in the early 2000s, it no longer mattered if an ICT start-up catered for a thousand, million, or billion users. This "infra-relativism" led to a culture of global indifference. The question no longer was whether or not these services could be delivered, but who got there first to secure the "lock-in" in order to establish the necessary monopoly. This is what platforms do: they do not create but eliminate markets.[6]

Let's go back in time and ask ourselves how we got here. Take S. Alexander Reed's *Assimilate*, which presents itself as "a critical history of industrial music."[7] Reed's account can be used as a mirror, an inspiration to tell the story of the 1990s "short summer" of network counter-culture, an avant-garde that was all too aware of its own post-1989 inability to make larger claims, let alone be utopian. Reed traces the "pure darkness" of industrial music back to Italian futurism, Artaud's Theatre of Cruelty, and William Burroughs' cut-up techniques. The

---

5　See Dick Hebdige, *Hiding in the Light: On Images and Things* (London: Routledge, 1989).

6　One of the first authors to describe this dotcom logic is Michael Wolff's *Burn Rate: How I Survived the Gold Rush Years on the Internet* (New York: Simon & Schuster, 1998). Wolff describes the venture capital logic as an aristocracy principle: it's all about land, not trade. The start-ups depend on the capital market rather than customer-based income. "We're an industry without income." Twenty years later this logic is still in place.

7　S. Alexander Reed, *Assimilate: A Critical History of Industrial Music* (Oxford: Oxford University Press, 2013).

sound of the squats coming out of the rust belts and deserted inner cities not only expressed the existential anger of a lost post-punk generation, it also produced early digital culture. This self-destructive Reagan/Thatcher era also transfigures into the first generation of personal computers that were used to produce zines and sound samples. Reed tells the story of isolated, self-producing small units. These "UFOs," as Patrick Codenys of Front 242 calls them, were autonomous nodes with a strong desire to communicate. According to Reed the isolation in this pre-internet period was "merely a geographic one: a vital connection exists between early industrial music and the global network established through the Fluxus art movement, its outgrowth of mail art, and the cassette, and small press cultures that arose in the late 1970s."[8] It is this cultural ecology, defined by weak ties of like-minded producers, that would be the ideal context in which the early internet could spread like a wild fire.

Surrounded by the doom and gloom of the neoliberal order with its permanent austerity, factory closures, the take-over of global finance, environmental disasters (from acid rain to Chernobyl), and mass unemployment, it was both tempting and subversive to embrace "the new" that the baby-boomer, post-war generation, and the powers-to-be had no clue about. Reed refers to musician La Monte Young's preference of the new over the good: "The new is a non-directional, non-teleological one, thus differing from the traditionalist and reactionary preconceptions of 'progress,' which were synonymous with 'good'."[9] "Good" was the realm of priests and politicians, academics, critics, and curators, and their conservative judgement had been predictable for years. Chaos and mess was not their preferred structure of feeling. According to the discourse police, the DIY aesthetics of the "ingenious dilettantes" was neither "professional" nor "pop" and was thus ignored. Networks were not good; for certain they were new, and yet invisible for authorities.

Much like the industrial music scene, early cyber-culture was ambivalent about its own democratic imperative. Networking was first and foremost networks-for-us. The claim to provide "access for all" (the infamous name of the Dutch hackers ISP that would be sold in 1998 to the former national telecom firm KPN) only came later and was an explicit counter-historical anomaly in an era when public utilities

---

8   Ibid., 111.
9   Ibid., 41.

were being carved up and privatized. Autonomy became synonymous with an inward-looking worldview one step away from total narcissism. Network tools had to be democratized. The concepts and software were easy to copy and install. The networks themselves were not necessarily open to all. If you got the groove, it was easy to find your way in.

Reed sums up this position accurately under the term "techno-ambivalence." "In his 1992 collaboration with the band Ministry, Burroughs orders us to 'Cut word lines. Cut music lines. Smash the control images. Smash the control machines.' This cutting and smashing is by no means a rejection outright of the viral agents of mind control – words, technology, and belief – but instead it's a reversal of these agent's powers upon themselves. As both the fragmented recordings to be cut up and as the recording device, machines are necessary to smash the machine, just as the vaccination is achieved through viral exposure."[10] The ambivalence between technophobe elements (computer as the 1984 control machine) and technophile (liberating production) promises remains unsolved. Take SPK's song *Metal Dance*, in which, according to Reed, the band attempts to have its revolution and dance to it too (in comparison to the now lame and politically correct warning that merely demobilizes collective desire: "If I Can't Dance, I Don't Want To Be Part of Your Revolution").

In the cultural context of the 1990s networks were neither inhabited by individuals (users with a "profile") nor by institutions. They created light and fluid swarms, not homogeneous masses. Consider networks as connectors between pockets of initiatives. Networks were not NGOs, neither did they have much resemblance with the emerging hipster start-ups. If any philosophy could come close in describing them, it would be the rhizomatic dreamscapes of Gilles Deleuze and Félix Guattari (ecstasy), combined with the leather jacket power politics of Michel Foucault (speed).[11] Without exception the "new media" collectives were products of previous social movements (squatting, feminism, ecology, anti-racism) and cannot be understood

---

10   Ibid., 40.

11   See Gilles Deleuze and Félix Guattari, *A Thousand Plateaus: Capitalism and Schizophrenia*, trans. B. Massumi, (Minneapolis: University of Minnesota Press, 1987) and Michel Foucault, *Foucault Live: Collected Interviews, 1961 – 1984*, edited by Sylvère Lotringer, trans. Lysa Hochroth and John Johnston (New York: Semiotext(e), 1989).

outside of that context. Following Adilkno's definition of "the movement as the memory of the event," there's a task here to reconstruct the origins of network initiatives.[12] What was their event in the past and what's the event today? It's too easy to say it ought to be located in the offline world. Certainly the social element is key, but not the question of whether the magic moment happened in real life or was mediated through machines. There was no need to make a distinction between the two.

Post-Facebook, the question is no longer about scale. No matter how much we all wish to have our fair share of exposure, networks can only scale down from here. That's when organized networks come into play. Orgnets are an organizational model that addresses institutional, technical, and political realities of the present. Global connectivity reached its moment of entropy some time in the period after September 11, the ongoing wars in the Middle East, the ubiquity of so-called Web 2.0, and the general consensus that humanity and capitalism have destroyed the planet. Tactics have now shifted to "meme design" inside a protected environment. Orgnets is an "out of season" concept because its time has either not yet come, already passed, or never materialized. In retrospect, we could claim that underground cultural networks dealing with industrial music, raves, zines, squatting, and independent publishing during the late 1980s and early 1990s had orgnet characteristics: the actors developed strong ties, despite the fact that they did not know each other and had to work across large distances. As today's social media platforms systematically neglect (read: ban) collective networking tools, local and regional organizations still have unknown revolutionary potentials, beyond the existing organizational forms such as the political party and event-based occupations and other forms of protest. For some, their decisive moment will be an image burnout. For others it will be war, permanent stagnation (or permanent vacation, as it was once called). What some fear as a "balkanization" of the net, many will celebrate as a true cultural, organizational, and eventually economic empowerment.

After the Party

Ultimately, network theory didn't go anywhere. Its normative approach in favor of the distributed network model rendered an entire

---

12   Adilkno, *Media Archive* (Brooklyn: Autonomedia, 1998), 16.

field-in-the-making irrelevant once rhizomes were replaced by scale-free platforms for the billions. Who still creates networks? Computers are supposed to do that for us. Companies and other authorities visualize and utilize our real existing networks for their purposes. We merely swipe, click, and like. What's left are network visualizations that no one seems to be able to read, not even the machines. Maps of network topologies are essentially eye candy, generated for the few networks in need of aesthetic affirmation. From an organizational perspective the network has not delivered either. It may be promising that one day vagueness and non-commitment might transform into firm, long-term engagement. But who's honestly going to wait for all these hyper-informed social media users that have no clue anymore about the basics of self-organization?

Jodi Dean's critique of the network form is interesting in this context. Her plea to return to the (communist) party reads like a Hegelian proposal to overcome dispersed short-term commitment. In *Crowds and Party* she asks how do mass protests become an organized activist collective? "How can acts remain intelligible as acts of a collective subject? How do people prevent their acts from being absorbed back into communicative capitalism?"[13] Social media architectures actively prevent autonomous organization (not to mention the obvious techniques of surveillance and aspects of social control). The "leaderless" Occupy approach was only able to orchestrate one-off protests and failed to set up sustainable grass-roots initiatives. Following in the footsteps of Elias Canetti, Dean states that "the crowd wants to endure," and pushes this desire in a particular direction by declaring that "the party provides an apparatus for this endurance."

According to Dean, what's missing in our current understanding is the "affective infrastructure of the party, its reconfiguration of the crowd unconsciousness into a political form." The party is presented as "the bearer of the lessons of the uprising." For Dean "the party, especially the communist party, operates as a transferential object – a symbol and combination of rituals and processes – for the collective action of the many." It is all about reconfigurations and reverberations, or overtone. "The party is tasked with transmitting the event's overtone." Regrettably this is a stillborn academic exercise as it presumes that Lenin is going to be a role model for the social media masses.

---

13   Jodi Dean, *Crowds and Party* (New York: Verso, 2016), 218.

Some might adopt his goatee beard as they guzzle down another latte, but that's about the extent of it. Why this self-defeating proposition to return to the party form celebrated by Marxist-Leninism is made remains unclear. The historical culmination of such an organizational form manifests as a socialist state that is structurally tied to capitalism. So really what's the substantive difference going on here for Dean and her intellectual inspirations and fellow-travelers such as Slavoj Žižek and Alain Badiou? There are no reports included in their various tracts, manifestos, and books of attempts to start such a party, or how join one. This makes the enterprise rather hollow, despite Dean and her cadre advocating the founding of the communist party for a number of years.

We see the challenge elsewhere, earlier in the process. A progressive meme design will have to start from scratch, developed and promoted in a protected yet participatory culture with the aim to beat the alt-right imaginary. Whether these motives, images, and role models will be used later on by a party remains to be seen. What also needs to be addressed is Dean's proposed transformative act of becoming a member as a way to ferment desire, to capture the energy of the collective event. Is it true that we all long to sign up and feel nostalgic about "membership"? There might be regression everywhere today, yet there are no signs for a "return" to membership organizations. We've all read the statistics of membership decline in unions, sports clubs, and religious organizations in the West. The social media ideology does not address us as committed members, we're merely users with a profile. How can we alter, and differentiate this dominant form of digital subjectivity?

As Jodi Dean rightly observes, the party form is no longer recognized as an affective infrastructure that can address problems. The 21st century political party is precisely not a form of concentration and endurance. The question shouldn't be party or no party. What's on the table is the strategic question regarding what the institutional form of this era will look like (presumably we want to reverse the current social entropy). The problem of Dean's approach is not one of analysis or urgency but one of over-determination. The question "what is to be done?" should be an open ended one. Agreed, we need synchronous political socialization, one that can overcome the feeling of being stuck in the lonely social media crowd. Let's see it as a start. The key to the problem lies elsewhere. It is "social networking" (as it is still called in Italy, rather than social media) that should be transformed. Let's not repeat the mistakes of the 90s cyber-generation who were utterly unprepared for the take-over by intermediaries such as Google, Amazon, and Facebook or, for that matter, Alibaba, Renren, and Weibo. We need contradictory platforms that break through the unconscious numbness of smooth interfaces. Let's build a toolkit and hack the attention economy. It should be easy to smash the online self and its boring cult of narcissism. These are the post-network challenges.

New Institutional Forms as Vehicles of Transition

If there is a legacy of the 20th century that might be worth looking into it will be neither communism nor the Party but "commoning" as a new form of organization. How can we shape the elements that we have in common into an organized form? Attempts initiated by the group around Michel Bauwens and the Peer-to-Peer Foundation or the "platform cooperativism" of Trebor Scholz and others demonstrate that sustainable networks are viable as long as you stick to the topic and build a movement together with a dedicated group.[14]

We consider the question of organizational form as central to a politics unhinged from the monopoly effects of platform capitalism. How to organize is always a question of media and mediation. Dean shows us the difficulty of supposing that political forms – whether as a party or otherwise – might somehow be distinct from media forms. The practice and concept of the "people's mic" suggests that even in the seemingly all too human moment of the general assembly, the capacity to amplify and relay sound across space is predicated on the repetition of bodies in machinic ways.

Political subjectivities are conditioned be media of operation. The possibility of digital media technologies and infrastructures constituting new social-political forms is not without its own challenges. British media scholar, Nick Couldry, asks: "What are the chances of creating new political *institutions* with sufficient authority to transform regimes of evaluation and challenge the framing of political

---

14   See P2P Foundation, HTTPS://P2PFOUNDATION.NET/ and Trebor Scholz,
     Uberworked and Underpaid: How Workers are Disrupting the Digital Economy
     (Cambridge: Polity Press, 2017).

space? I suggest they are small: if well-established political institutions' possibilities of 'sustained performance across events and issues' becomes more difficult, how much more difficult is it to establish new political institutions with the authority required for sustained programmes of radical policy action?"[15] Where Couldry's focus is on how political space is framed and evaluated in ways that command and sustain authority, we find the emphasis on chance here as one beholden to a particular political mindset, however latent that may be.

Chance is the foundation of technocratic game theory. What, for instance, were the chances of the internet in the mid-eighties? Such a question belies a form of neo-conservatism, hidden in statistics. What was the chance of a revolution in Russia, early 1917? Systems implode, and Couldry is obviously not ready for that. Markets crash. As do ruling political institutions. The question is, who's ready to take over? Baudrillard was right: we long for an explosion, and all we got is a lousy implosion, a never-ending stagnation, Japanese style (1980s predictions were correct, Japan is the 21st century role model, but a rather different one from what was predicted).

Who's got a plan? Over the past decade the geopolitical shift to global markets and centers in East Asia has impacted enormously on the economic and social fabric enjoyed in North America and Europe for a few decades following World War II. With new technologies of automation now impacting employment prospects across the world, what happens when 20%, 40%, 60% of the population is written off, without a job, and sliding into a life of destitution below the poverty line? Democracy as an orchestrated ensemble of the elites falls apart. Even the seeming stability of authoritarian capitalism in countries like China will rapidly struggle to govern populations in conditions of mass crisis.

The creation of new institutions will only happen once the old ones have gone. Foucault's criticism of revolution was that inevitably the new guard simply end up occupying the warmed up seats of the old guard. "In order to be able to fight a State which is more than just a government, the revolutionary movement must possess equivalent politico-military forces and hence must constitute itself as a party, organized internally in the same way as a State apparatus

---

15   Nick Couldry, *Media, Society, World: Social Theory and Digital Media Practice* (Cambridge: Polity Press, 2012).

with the same mechanisms of hierarchies and organization of powers. This consequence is heavy with significance."[16] While an element of structural determinism lurks within Foucault's response to his Marxist interlocutors, his statement nonetheless invites the question: what is the difference between revolution (as a reproduction of the same) and taking control of the infrastructures of those in power? Neither result in an invention of new institutional forms. When movements organize as a party the possibility of alternatives is extinguished. This is the brilliance of Foucault's analysis, and a position that Dean reproduces in her valorization of the party as the primary vehicle for political articulation. In both cases, however, there is nowhere left for radical politics within organizational apparatuses of equivalence.

There's a legitimacy crisis for new institutions. The New can also reach a crisis status well before it has come to the level of full implementation. The imaginary and real power of existing institutional frameworks so often work against the possibility of new institutional forms arising with the capacity to displace existing powers. In this regard, it is more strategic to consider orgnets as transitional vehicles. They are not the solution, but not the problem either. That much is clear. The proposal and push by ex-minister of finance in Greece's Syriza government, Yanis Varoufakis, to create a pan-European network, DiEM25, is similarly caught within the trap of solutionism.[17] DiEM is not an "alternative for the European Commission." That's nonsense. DiEM is neither a pan-European think tank, nor an NGO. If anything it is a networked movement. It creates European, national, and local networks. DiEM prepares people, but in the event of an EU collapse DiEM is not going to be less ugly because of it. With "a view to conjuring up a democratic surge across Europe, a common European identity, an authentic European sovereignty, an internationalist bulwark against both submission to Brussels and hyper-nationalist reaction," the utopianism of DiEM transfers the core tenets of liberal democracy from the sovereign power of nation-states to the populist delusion of networks able to govern populations in scale-free

---

16 Michel Foucault, 'Body/Power', in *Power/Knowledge: Selected Interviews and Other Writings 1972-1977*, edited by Colin Gordon, trans. Colin Gordon, Leo Marshall, John Mepham, and Kate Soper (New York: Pantheon Books, 1972), 59.

17 DiEM25 Manifesto, HTTPS://DIEM25.ORG/MANIFESTO-LONG/.

ways. Any organizational entity founded on the values espoused by DiEM is necessarily an entity that traffics in the politics of exclusion, which, like Weber's concept of the modern state, is predicated on a "monopoly of violence."[18] DiEM is not in any rush to point out the histories of colonialism that in so many ways condition the world of migration and lifestyles of contemporary Europe. DiEM's common Europe of transparent decision making is not open to barbarians beyond its borders and instead focuses on inner-European inequalities. Varoufakis is not prepared to reconcile this logical endpoint of a post-EU world with networks that function as *de facto* states.[19]

Needless to say, DiEM can be conceived as transitional vehicle or messenger that might support radical policy reforms within the Brussels lobby scene. But such a status can often enough precipitate an identity crisis, and this is exactly what has happened with DiEM (no matter that it's unwillingly to acknowledge this, at least in public). DiEM claims to be a network of movements from below while also functioning much like a political party where meet-ups all too often are assumed as equivalent to (representational) membership. We recall a meeting between DiEM and political party Die Linke (The Left) in September 2016 held at Astra-Kulturhaus, an indie rock venue in Berlin's hip district of Friedrichshain. In between a rush of other appointments, Varoufakis took to the stage with all of the ease and frothy eroticism expected of political rock stars these days (Corbyn and Sanders aside). In his message to comrades, Varoufakis made a point of highlighting a White Paper recently prepared by the DiEM executive (or "coordinating collective") for tabling in Brussels. Stunning here was the assumption that a document as deadening as the genre of policy recommendations might somehow do the magical work of galvanizing movements into action, let alone sustain political passions. Maybe that can happen inside political party headquarters, but it's highly unlikely amongst social movements. Nor, for that matter, was anyone in Brussels about to lend credibility to a report coming from DiEM.

---

18 Max Weber, "Politics as Vocation," in Max Weber, *Complete Writings on Academic and Political Vocations*, ed. John Dreijmanis, trans. Gordon C. Wells (New York: Agora Publishing, 2008), 155–207.

19 See Yanis Varoufakis, "Why We Must Save the EU," *The Guardian*, April 5, 2016, https://www.theguardian.com/world/2016/apr/05/yanis-varoufakis-why-we-must-save-the-eu.

So, is DiEM25 a movement or party? Or perhaps a new entity altogether? Reporting in *Il Manifesto* on an earlier meeting in Berlin in February 2016 for the launch of the DiEM 2025 manifesto, Marco Bascetta and Sandro Mezzadra signal the inherent contradiction when an embryonic movement traffics generic notions of democracy that also have to scale to the supra-state level of the EU.[20] As many political theorists and philosophers are want to say, democracy is an indefinite project, always to be deferred. As an aspiration of a society to come, perhaps the key problem facing DiEM is the confusion it has at the level of organizational form. DiEM is neither a movement or a party, yet it cannot help but try and act as both. This predicament is one shared with Dean, who also imagines a continuum can be stretched from the assemblies to the party form. Podemos and the earlier incarnation of Syriza have perhaps more than others been able to straddle the tension of populist politics caught between the party and the street. But in both cases the movements eventually drift away, returning to their former fragmented sects. The impasse of democracy as an imaginary and desire around which politics is organized might better be put aside for a politics of struggle that focuses instead on other names.

Organizing Next Nature

Let's forget about organized platforms (that's a monstrous contradiction) and bet on a necessary renaissance of the network mode. What role might orgnets play in the core debates around environmental catastrophe that define the increasingly rapid decimation of planetary life as we know it? Can a social-technical mode of doing things in collective ways in the world have any correlation with techno-ecologies populated by robots, automated systems (AI, machine learning, enterprise resource planning software), and extractive machines? The question of organization will never go away. The social can be calculated, mapped, simulated, and ultimately eliminated, but the act of organizing itself can never fully be outsourced. No matter how much economy, labor, and life are defined by automation and repetition, the

---

20   Marco Bascetta and Sandro Mezzadra, "La costituente di Varoufakis sociale e non sovranista," *Il Manifesto*, February 9, 2016. Rough English translation at: HTTPS://GLOBAL.ILMANIFESTO.IT/ VAROUFAKIS-APPEALS-FOR-DEMOCRATIC-AWAKENING/.

machine needs constant maintenance, resources, upgrades, supervision, care. The substrate from which organization emerges – whether this is human, machine, environment, or thing – is first and foremost a relation of transformation underscored by material, affective, social, and kinetic propensities.

For German media philosopher Erich Hörl, the "general ecology" of the techno-sphere analyses the contemporary condition of governance and cybernetic control in a technical world. Hörl maintains we are in an "environmental culture of control that, thanks to the radical environmental distribution of agency by environmental media technologies, ranging from sensorial to algorithmic environments, from bio- to nano- and geo-technologies, renders environmentality visible and prioritizes it like never before."[21] Yet environmentality understood as a new idiom of control is only visible in as much as it manifests on a scale of perceptible transformation. The infrastructural and technical components of environmental media are more often highly secluded and inaccessible data facilities, or computational systems operating in the background of routine transactions, processes, and practices. The political question of power goes beyond a philosophical politics of sense, theory, and concepts.[22] To attribute a politics to such struggles of thought we would need to identify the institutional and geocultural terrains in which conceptual dispute is materialized.

We agree with Hörl that a techno-environmentality paradigm succeeds and displaces the primacy of human agency and bind of reason. There's an embarrassing juvenility that attends the human pretense of control. Though we would sideline the question of politics as a problem for theory ("decision design") and instead ask how environmental media relates to the organization and politics of movements. In terms of a program for orgnets operating within these sort of parameters, one critical question concerns how to organize in ways that are responsive to new infrastructures of distribution and new agents of power? A techno-ecology of robots and automation receives a steady stream of reporting in the mainstream press and tech-magazines. The eradication of jobs is the common narrative across these reports. The

---

21   Erich Hörl, "Introduction to General Ecology," in Erich Hörl with James Burton (eds), *General Ecology: A New Ecological Paradigm* (London: Bloomsbury Academic, 2017), 9.

22   See Hörl, 5, 14.

displacement of the human as the primary agent of change in the world is thus coincident with the increasing extension of technical environments that manage social and economic life. Why don't we switch our attention instead to architectures of inoperability? One tiny (unknown) disruption and the robot falls silent – that's the new certainty of our age, where "the 'assembly life' [has] replaced the assembly line."[23]

Another case could be the Next Nature Network, an Amsterdam cultural organization aka non-profit design collective that sets out to draw up scenarios in which the society of adaptation is amplified in ways that suggest new species-beings will unsettle the established order of things. Such a program resonates with Hörl's general ecology. Inquiring into the techno-environments that condition the adaptation of life in an epoch of mass distinction is at the core of the project by the Next Nature Network. Their question of organization is, unlike Hörl, also a social-political one. Among the many "unlikely futures" that Next Nature conceptualizes in its blog postings, presentations, books, and exhibitions, it is their proposal for a radical remake of the (Dutch) ecology movement that interests us most here. How can green activism shed its outdated, romantic, 19th century version of nature and develop a new understanding of politics that integrates human interventions into a version of nature as a radical design? "Virtual worlds, printed food, living cities and wild robots; we're so surrounded by technology that it's becoming our next nature. Next Nature Network is *the* international network for anyone interested to join the debate on our future – in which nature and technology are fusing."[24] Their slogan: Forward to Nature! Heritage and conservation are deconstructed from a post-human perspective, becoming legacy architectures and systems that condition contemporary transformations. But what's even more interesting about this "prototyping" of protest is the demand that the "general ecology" itself should be included in the design of future forms of organizing. The aim of NNN is topical: how to transform a (cultural) organization into a movement, using the network as a vehicle?

What do orgnets mean within such a context? Welcome to the social stack. How can we think of orgnets as the contemporary expression of

23   Sylvère Lotringer, "Better than Life," *Artforum International* 41 (April, 2003): 194–97, 252–53.

24   Next Nature Network, HTTPS://WWW.NEXTNATURE.NET/WELCOME/.

collective human activity? Can they operate as a transitionary vehicle in this period of techno-ecological consolidation? Should orgnets be seen as organizational forms that facilitate alternative futures, producing unforeseen mixtures of critique and innovation, or are they better placed to help realize the seeming inevitability of a world ruled by machines? The latter is the accelerationist option that celebrates the nihilism of capital accumulation and an anticipatory delirium of the revolutionary event. The former suggests a reformist agenda – a position for us that holds minimal appeal and is all too often beholden to a moral code of progressive politics predicated on identity and exclusion.

In our times of economic stagnation and ideological austerity, orgnets are a model to combat the hyper-efficient distribution of poverty. Stop sharing, start organizing. The orgnets question is first and foremost a media question regarding the organizational logic of technological forms and instituent practices. This means it lends itself to a critique of control, which at the current conjuncture is emerging and indeed consolidating in the form of platform capitalism. If we focus on the organizational logic of instituent practices, then we need to address the social dimension that creates alternatives to the monopoly effects of platform capitalism. To live a life outside the partitioned walls of platform capitalism and social media we really do have to accept the fact that no one will hand out better solutions on a plate. There is no technological fix. We have become deeply enamored with the so-called "free services" of platform capitalism and all too willing to open our data-generating selves to inspection and extraction economies. Sure, one option is to sit it out and wait for the demise of platform capitalism. That will happen. But meanwhile life passes by and models of organizing society and economy remain in the hands of the few (and they usually are nasty by default).

How to instigate a federated culture of networks combined with user-friendly secure communication is, for us, key to the collective design of a general ecology of our techno-spheres. This is not the fantasy of interoperability espoused by logistical media industries and operators jostling like platform capitalists for market control, but rather a loose alliance of strong ties that comes together out of collective desires for autonomous production. Inspiring on this front is the transit-organization work of the Barcelona Initiative of Technological Sovereignty (BITS).[25]

---

25   https://bits.city/.

Drawing on the history and experience of organizing local councils in southern Europe (particularly in Italy and Spain), BITS brings activists from the movements together with academics and policy makers to reorganize the distribution of services and support within urban settings. Such initiatives actively produce autonomous infrastructures of distribution and have no interest in passively adopting the free solutions dished out by Silicon Valley. Connected to the spread of Right to the City and Rebel City movements (Lefebvre, Harvey), but drawn from much longer histories of self-determination and localized critiques of urban renewal, the strategic question to draw from these meso-political networks is how to design scalability into these local efforts in order to address the politics of distribution.[26] We are not thinking here of scale as in scale-free networks, but rather scale as a technique through which social and political relations are forged to address particular problems that are often very local.

Organization is inseparable from experimentation and design. Platform capitalism demonstrates loud and clear that neither of these are bundled into the free software of service providers. The regressive propositions for a revived party politics are also not going to do the job. Movements are made when passions are not only ignited, but also organized in ways that respond to our media situation.

---

26 See Henri Lefebvre, "The Right to the City," in *Writings on Cities*, eds. and trans. Eleonore Kofman and Elizabeth Lebas (Cambridge, Mass.: Blackwell Publishers, 1996), 147–59 and David Harvey, *Rebel Cities: From the Right to the City to the Urban Revolution* (London: Verso, 2012).

BIBLIOGRAPHY

RATHER THAN REPRODUCE ALL OF THE REFERENCES LISTED IN OUR footnotes, we have instead selected here a range of readings that we find inspirational or serve as points of entry into writing about organized networks.

Adilkno. *Media Archive* (Brooklyn: Autonomedia, 1998).

autonomous a.f.r.i.k.a. gruppe (aka Luther Blissett and Sonja Brünzels). *Handbuch der Kommunikations Guerilla* (Berlin: Verlag Libertäre Assoziation, 1997).

Bakunin, Michael. *Statism and Anarchy* (Cambridge: Cambridge University Press, 1990).

Bakunin, Mikhail. *The Basic Bakunin Writings 1869-1871*, edited and trans. by Robert M. Cutler (Amherst: Prometheus Books, 1992).

Baudrillard, Jean. *In the Shadow of the Silent Majorities . . . Or the End of the Social and Other Essays*, trans. Paul Foss, Paul Patton, and John Johnson (New York: Semiotext(e), 1983).

Berardi, Franco "Bifo." *Heroes: Mass Murder and Suicide* (London: Verso, 2015).

Bratton, Benjamin. *The Stack: On Software and Sovereignty* (Cambridge, Mass.: MIT Press, 2015).

Coleman, Gabriella. *Hacker, Hoaxer, Whistleblower, Spy: The Many Faces of Anonymous* (London and New York: Verso, 2014).

Dean, Jodi. *Crowds and Party* (New York: Verso, 2016).

Doctorow, Cory. *For the Win* (New York: Tom Doherty Associates, 2010).

Easterling, Keller. *Enduring Innocence: Global Architecture and its Political Masquerades* (Cambridge, Mass.: MIT Press, 2005).

Fisher, Mark. *Ghosts of My Life: Writings on Depression, Hauntology, and Lost Futures* (Winchester: Zero Books, 2014).

Flusser, Vilém. *Post-History*, trans. Rodrigo Maltez Novaes (Minneapolis: Univocal Publishing, 2013 [1983]).

Guattari, Félix. *Psychoanalysis and Transversality: Texts and Interviews, 1955–1971*, trans. Ames Hodges (South Pasadena: Semiotext(e), 2015 [1972]).

Holmes, Brian. *Escape the Overcode: Activist Art in the Control Society* (Eindhoven and Zagreb: Van Abbemuseum Public Research / WHW, 2009).

Horkheimer, Max. *Dawn & Decline: Notes, 1926–1931 & 1950–1969*, trans. Michael Shaw (New York: Seabury Press, 1978).

Innis, Harold A. *The Bias of Communication* (Toronto: University of Toronto Press, 1951).

Kittler, Friedrich A. *Literature, Media, Information Systems: Essays* (Amsterdam: G + B Arts International, 1997).

McCarthy, Mary. *The Group* (London: Weidenfeld and Nicolson, 1963).

Martin, Reinhold. *The Organizational Complex: Architecture, Media, and Corporate Space* (Cambridge, Mass: MIT Press, 2003).

Montaigne, Michel de. *The Complete works of Montaigne: Essays, Travel Journal, Letters*, trans. Donald M. Frame (Stanford: Stanford University Press, 1957).

Mouffe, Chantal. *Agonistics: Thinking the World Politically* (London and New York Verso Books, 2013).

Moulier Boutang, Yann. *Cognitive Capitalism*, trans. Ed Emery (Cambridge: Polity Press, 2011 [2007]).

Munster, Anna. *An Aesthesia of Networks: Conjunctive Experience in Art and Technology* (Cambridge, Mass.: MIT Press, 2013).

Pickering, Andrew. *The Cybernetic Brain: Sketches of Another Future* (Chicago: University of Chicago Press, 2010).

Scholz, Trebor. *Uberworked and Underpaid: How Workers are Disrupting the Digital Economy* (Cambridge: Polity Press, 2017).

Srnicek, Nick. *Platform Capitalism* (Cambridge: Polity Press, 2017).

Terranova, Tiziana. "Red Stack Attack! Algorithms, Capital, and the Automation of the Common," in Robin Mackay and Armen Avanessian (eds), *#Accelerate: The Accelerationist Reader* (Falmouth: Urbanomic, 2014), 381–99.

The Invisible Committee. *The Coming Insurrection* (South Pasadena: Semiotext(e), 2009).

The Invisible Committee. *To Our Friends*, trans. Robert Hurley (South Pasadena: Semiotext(e), 2014).

Stiegler, Bernard. *Taking Care of Youth and the Generations*, trans. Stephen Barker (Stanford: Stanford University Press, 2010).

Whyte, William H. *The Organization Man* (Philadelphia: University of Pennsylvania Press, 2002 [1956]).

The texts in this volume were previously published in the following projects and collections:

"Tactical Media and the Question of Organization," in Eric Kluitenberg and David Garcia (eds), *Tactical Media Anthology* (Cambridge, Mass.: MIT Press, in press).

"Organization in Platform Capitalism," in Rosi Braidotti and Maria Hlavajova (eds), *Posthuman Glossary* (London: Bloomsbury, 2018), 306–8.

"Organized Networks in the Age of Platform Capitalism," in Graham Miekle (ed.), *The Routledge Companion to Media and Activism* (Abingdon: Routledge, 2018), 404–13.

"The Politics of Organized Networks: The Art of Collective Coordination and the Seriality of Demands," in Wendy Hui Kyong Chun, Anna Watkins Fisher, and Thomas Keenan (eds), *New Media, Old Media: A History and Theory Reader*, 2nd edition (New York: Routledge, 2016), 335–45.

"Network Cultures and the Architecture of Decision," in Lina Dencik and Oliver Leistert (eds), *Critical Perspectives on Social Media and Protest: Between Control and Emancipation* (New York: Rowman & Littlefield, 2015), 219–32.

"Concentrate your Power: Design Organized Networks," in *Future Mirrors* (Sydney: Modelab, 2015), 31–33, HTTPS://WWW. MODELAB.INFO/FUTURE-MIRRORS

"Organizing Networks," in Steirischer Herbst and Florian Malzacher (eds), *Truth is Concrete: A Handbook for Artistic Strategies in Real Politics* (Berlin: Sternberg Press, 2014), 144–46.

"Organised Networks: Weak Ties to Strong Links," *Occupied Times* 23 (November, 2013), HTTP://THEOCCUPIEDTIMES.ORG/?P=12358

"In Praise of Concept Production: Formats, Schools and Non-Representational Media Studies," in Kelly Gates (ed.), *Media Studies Futures, The International Encyclopedia of Media Studies, Vol. 5* (Cambridge and Malden, Mass.: Wiley-Blackwell, 2013), 61–75.

"'Seriality for All': The Role of Protocols and Standards in Critical Theory," in Pieter Wisse (ed.), *Interoperabel Nederland* (The Hague: Forum Standaardisatie, 2011), 426–33.

"Understanding Cartopolitics: The Logic of Networks, From Visualisation to Organisation," in Geraldine Barlow (ed.), *Networks (cells & silos)* (Melbourne: Monash University Museum of Art, 2011), 29–31.

"Understanding Cartopolitics: The Logic of Networks, From Visualisation to Organisation," *Acoustic Space: Journal for Transdisciplinary Research of Art, Science, Technology and Society* 10 (2011): 53–56, 57–59. [Special issue on Networks and Sustainability]. (Trans. English and Latvian)

"Urgent Aphorisms: Notes on Organized Networks for the Connected Multitudes," in Mark Deuze (ed.), *Managing Media Work* (London: Sage, 2011), 279–90.

"The Portable Guide to Organized Networks," *Networks and Sustainability*, Nice Paper no. 3 (Riga: RIXC, The Centre for New Media Culture, 2010), 8–9.

"Ten Theses on Non-Democratic Electronics: Organized Networks Updated," in Marco Berlinguer and Hilary Wainwright (eds), *Networked Politics: Rethinking Political Organisation in an Age of Movements and Networks* (Berlin: Rosa Luxemburg Foundation, 2007), 61–65.

"Dawn of the Organised Networks," *Fibreculture Journal* 5 (2005), HTTP://FIVE.FIBRECULTUREJOURNAL.ORG/ FCJ-029-DAWN-OF-THE-ORGANISED-NETWORKS

MINOR COMPOSITIONS

Occupation Culture – Alan W. Moore
Crisis to Insurrection – Mikkel Bolt Rasmussen
Gee Vaucher. Introspective – Ed. Stevphen Shukaitis
The Aesthetic of Our Anger – Ed. Mike Dines & Matthew Worley
The Way Out – Kasper Opstrup
Situating Ourselves in Displacement – Ed. Marc Herbst et al.